This is a guide to the use of the most popular type of telescope in the world, the 20-cm (8-inch) Schmidt–Cassegrain telescope. This compact instrument revolutionized amateur astronomy and astro-photography, and more than ten thousand are purchased each year. Peter Manly, a devotee and keen user of the Schmidt–Cassegrain, takes the telescope owner through all aspects of using the telescope in easy stages. It starts with techniques for viewing the Moon, then takes the observer through our planetary system, and on to the deep sky, where nebulae and galaxies are treated extensively. There are interesting projects to try, such as observing the nearest star and chasing eclipses. A full range of telescope accessories and detectors is described together with advice on their use.

The 20-cm Schmidt–Cassegrain Telescope

THE 20-CM SCHMIDT–CASSEGRAIN TELESCOPE

Peter L. Manly
Syzygy+, Tempe, Arizona

CAMBRIDGE
UNIVERSITY PRESS

Published by the Press Syndicate of the University of Cambridge
The Pitt Building, Trumpington Street, Cambridge CB2 1RP
40 West 20th Street, New York, NY 10011–4211, USA
10 Stamford Road, Oakleigh, Melbourne 3166, Australia

First published 1994

Printed in Great Britain at the University Press, Cambridge

A catalogue for this book is available from the British Library

Library of Congress cataloguing in publication data
Manly, Peter L.
The 20-cm. Schmidt Cassegrain telescope / Peter L. Manly.
p.   cm.
Includes index.
ISBN 0 521 43360 6
1. Schmidt telescopes.   I. Title.   II. Title: Twenty-centimeter
Schmidt Cassegrain telescope.
QB88.M25   1994
522'.2–dc20   93-42045 CIP

ISBN 0 521 43360 6 hardback

Contents

CONTENTS

Illustrations

Preface

When Simon Mitton of Cambridge University Press first suggested this book at the 1991 Riverside Telescope Maker's Conference, I didn't think he was talking to me. For years I'd been primarily concerned with all manner of odd-ball optical devices from satellite tracking systems to airborne telescopes as a professional astronomer. Then it finally sank in that after I'd finished my employer's work I'd go home and drag my trusty little 20-cm Schmidt–Cassegrain (S–C) out in the back yard just to have fun. For a decade and a half it has been part of our family, affording many hours of pleasure and not a little education for my children who have progressed from barely crawling to college entrance exams in the interim.

I believe I saw my first catadioptric telescope[1] during preparations for the International Geophysical Year in the late 1950s. In those days only professionals could own such a glorious and expensive machine. I was disappointed that we students would have to use smaller telescopes to track the yet-to-be-launched first artificial Earth satellite. But I did get to look through the shiny blue and white painted 'professional' telescope at times and I vowed that one day I would have one for my very own. Two decades would elapse before I could purchase a used 20-cm S–C.[2]

[1] Catadioptric telescopes use both mirrors and lenses as opposed to refractors which use lenses only and reflectors which use mirrors only. The corrector plate at the front end of a Schmidt–Cassegrain (S–C) acts as a lens and thus the type falls under the general heading of a catadioptric telescope which also includes Maksutov telescopes and Schmidt cameras.

[2] In astronomy there exists a rift between telescope designers/builders and telescope observers. Telescope makers have been known to sneer at those who use "store bought" instruments. Often it is said that the only mark of intelligence is the ability to grind and polish (manually) a quarter wave mirror. Indeed, I have done so and it is an exhausting educational experience. On the other hand, it has been quipped that the mark of stupidity is to grind and polish a second mirror.

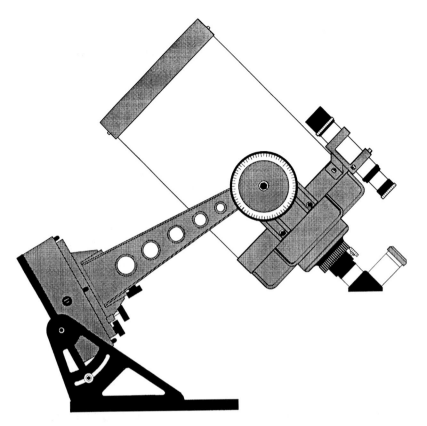

Figure 1. The classical Schmidt–Cassegrain

The aperture of a 20-cm diameter[3] telescope has an area of 314.16 cm^2. Any Schmidt–Cassegrain (S–C) telescope design, like all telescopes, is an engineering trade-off, applying sound design principles to the problem of observing the sky at a reasonable cost.[4] Newtonian telescopes generally have better wide-field performance at about half the price (per centimeter of aperture) of an S–C but they are physically bulky and not as easy to operate

[3] 20 cm equals just about 8 inches (7.874 015 748 inches, to be more exact) and for those readers who insist on using the archaic English measurement system, the aperture area is 4.87X10^{-9} square miles.

[4] The discussion of which telescope type is best has raged for many years. Typical is a set of articles by Harry D. Jamieson in the *Journal of the Association of Lunar and Planetary Observers* (ALPO) starting in Volume 35, Number 4, December, 1992, p. 181. Throughout the next four issues, letters debating the point appeared. At the time of writing I await the next issue which will probably continue the discussion.

as the S–C. Refractors are reputed to have better narrow field performance on planets and double stars (although a properly adjusted S–C usually challenges most refractors' performance at a fraction of the cost).

The S–C is a general purpose astronomical instrument, capable of superb views for amateur[5] observations and able to turn out professional results in the hands of an experienced astronomer. Like any optical and mechanical system, its design is a series of compromises between performance and costs. Commercial telescope manufacturers have the added pressures of maintaining high optical and mechanical quality while remaining competitive in price. The large commercial market for a good 20-cm S–C has forced the several manufacturers of the type to work hard at producing a high quality, affordable telescope.[6] Competition in the free market has thus reduced the cost of a 20-cm S–C from a 'professionals only' instrument to one which can be purchased by nearly any amateur.

This is not a comparison shopper's guide since I'm not going to tell you which manufacturer's telescope to buy for your observing program.[7] I haven't the foggiest idea of what your observing program is. And your program may shift emphasis as time passes. Don't worry that your astronomical interests may drift, causing you to buy some new type of telescope. The S–C is capable of handling a wide variety of observing programs. You can also modify your telescope and add accessories to handle may different types of observations. The second reason that I'm not going to recommend a specific telescope brand is that, with a little care, any competent manufacturer can make diffraction limited 20-cm diameter f/10 optics.[8] The

---

[5]   The word amateur comes from the Latin 'amator' which means 'lover'. Reference *The Cambridge Astronomy Guide*, by Bill Liller and Ben Mayer, Cambridge University Press, 1985, p. 9. Thus, an amateur observes because he loves the experience. A professional, by definition, is one who is paid to work. It has often been quipped that only the most dedicated professional astronomers can ever achieve the status of amateur — one who truly loves astronomy and who would observe without remuneration.

[6]   It is estimated that between 75 000 and 100 000 20-cm S–C telescopes have been made by the several manufacturers.

[7]   Readers familiar with 20 cm S–C telescopes may note that one particular manufacturer's model is shown in most illustrations. It is neither better nor less expensive than any competitor's models. That one just happened to be handy when I needed illustrations.

[8]   The diffraction limit of optical resolution is a function of the diameter of the aperture. Once the optical figure of the mirror is good enough that the glass is better than the diffraction limit, then any further polishing will not increase resolution. An increase in resolution will require an increase in the diameter of the whole telescope.

differences between scopes will be in the mounting, ruggedness, accessories, auxiliary fittings and, occasionally, optical coatings.

I will assume that you have read the telescope owner's manual. Maybe you haven't memorized it but you've run your eyes down the pages at least once. I'll assume you know the differences between planets, moons, asteroids, comets, stars and galaxies. Similarly, the system of measuring brightness in magnitudes and position in celestial coordinates should be understood, although they may be new and unfamiliar just now.[9]

The book is organized along the lines of learning something about your telescope and then applying it to a specific observation. While later chapters generally cover the finer points of telescope operation and more difficult observing tasks, you do not need to cover the chapters in numerical order, although it is recommended. The observing programs are designed to introduce the observer to many of the mundane objects in the Universe. In addition, special attention is paid to the more bizarre objects in the celestial zoo such as the brightest quasar, the star with the fastest proper motion[10] and (for Southern Hemisphere observers) the star which is closest to our own Solar system. We will examine a stellar nursery where stars are being born as we look at them and will look at a failed star which, if it had more of the right stuff, could shine like the Sun but as it is, just glows weakly in the infrared.

The book will discuss typical telescope performance with instruments operated by normal people at easily accessible observing sites. Occasionally, I will discuss the limit of performance but that refers to perhaps ten mountain peaks in the world on two or three nights of good seeing per year when operated by one of a handful of professional observers who are well

---

[9] There are various coordinate systems in use. This book shall concentrate on right ascension and declination, an Earth-oriented system, with an occasional excursion into altitude/azimuth coordinates when discussing mount alignment. Many astronomers use galactic coordinates, a spherical system oriented to the disk of our local galaxy. For those of you who intend to wander the galaxy while observing, write me and I can give you a handy little computer program to convert RA/Dec to galactic coordinates. If you intend to leave the home galaxy while observing, you're on your own.

[10] The proper motion of a star, asteroid or comet is its velocity North, South, East or West on the sky. It may also have an additional velocity component toward or away from us but that is referred to as its radial motion.

[11] Theoretically, a 20-cm telescope should allow an observer to see stars of Mv 14.2 according to ASTRONOMY Magazine, Nov. 1991, supplement, p. 9.

rested, breathing pure oxygen and have been dark adapted since last Tuesday.[11]

Just how faint an object can you expect to see with a 20-cm S–C? If your eyesight is average then maybe a twelfth or thirteenth magnitude star.[12] Galaxies, nebulae and comets are a little more difficult to estimate because their brightness is usually specified as the total amount of light which they emit but they're not point sources. Their visibility depends on their surface brightness and the eyepiece magnification which you use. It will also depend on the sky brightness at your observing site and your own visual acuity in discerning low surface brightness phenomena. You should at least be able to see all of the hundred or so objects in the Messier Catalogue with your 20-cm S–C telescope (providing, of course, that they rise above your local horizon).

If you've just purchased your 20-cm S–C and the sky is getting dark, I suspect you want to play with your new toy and you don't want to be sitting inside reading some book. Grab your owner's manual, review the parts about setting up your telescope and then head outdoors. Look at the Moon, find a planet. At least let the telescope see first light. Later, you'll learn some of the finer points of observing but for now, get out into the shadow of the telescope.

[12] The limiting sensitivity of a 20-cm telescope with one of the newer Charge Coupled Device (CCD) TV sensors is about Mv 17.5 although detection of Mv 20 stars has been reported. See the letter by Anthony Mallama in *Sky & Telescope Magazine*, February, 1993, p. 84.

Acknowledgements

Many people have helped in the preparation of this book but because it draws upon a couple of decades of observing with a 20-cm S–C, almost every astronomer I have ever met has contributed some small bit of knowledge.[13] Several individuals, however, have taken an active interest in the book. They include Mark Coco, Alan Hale and Tom Johnson of Celestron International, John Diebel and Scott Roberts of Meade Instruments Corporation, Raul Espinoza, Bruce Hoult, Gene Lucas, Chris Marriott, Gary Mussar, Joe Perry and Chris Schur.

This book was prepared using Apple Macintosh™ model SE and LC computers and Microsoft Word® software to create text. MacDraw™ and MacDraft™ software were used to create drawings. GIFConverter© software was used to import image files from other computer formats. Apple Hypercard™ and Multifinder™ software proved invaluable in organizing information and planning the production of the book. Finally, I would like to thank my copy-editor Beverley Lawrence, and my publisher for making it easy on writers by encouraging them to submit books on floppy disks.

---

[13] The first two astronomers I ever met were not professionals in the business. My Dad, a civil engineer by trade, dragged me out of bed one morning to observe a partial Solar eclipse and got me hooked on the sky. Mr Edwin Eide, a teacher, showed me that astronomy is an organized body of knowledge which is alive with researchers pushing at the frontiers of knowledge. He also taught me that this is a good working definition of Science.

1

History and development of the 20-cm Schmidt–Cassegrain

The Cassegrain telescope was invented in about 1672 by Sieur Guillaume Cassegrain, a sculptor and metal founder employed by the King of France.[14] The mirrors were probably made of speculum or some similar metal which could take a polish. Cassegrain communicated his design to Monsieur de Bercè who took it to the French Academy and by May of that year the idea had appeared in *Journal des Sçavans* and *Philosophical Transactions*. Almost immediately a dispute erupted concerning the relative qualities of the Cassegrain and two English telescope types, the newly invented Newtonian telescope and the older Gregorian design. Battle lines among scholars and their scientific academies were drawn along national boundaries – in this case down the middle of the English Channel. Even Sir Isaac Newton himself joined the fray and published dissertations establishing his claims.

The Gregorian telescope is composed of a concave parabolic primary mirror with a concave elliptical secondary. The Cassegrain uses a similar concave parabolic primary with a convex hyperbolic secondary. The Newtonian has a concave parabolic primary with a plane secondary. All three designs had the advantage of no color aberration since they used mirrors instead of lenses. The invention of the achromatic lens was a half century in the future at the time. The Gregorian design was popular because it produced an erect image and it was free from spherical aberration – theoretically. The technology was not available, however, for grinding accurate aspherical surfaces. The Cassegrain was a more compact design but it, too, required aspherical surfaces. Newton's telescope used only one curved surface, the other being an easily generated plane. Newton generally used a higher

---

[14] *The Telescope*, Louis Bell, Dover Publications, 1981 edition, p. 22. See also *The History of the Telescope*, Henry C. King, Dover Publications, 1955, p. 74, and *Lens Design Fundamentals*, Rudolf Kingslake, Academic Press, 1978, p. 322.

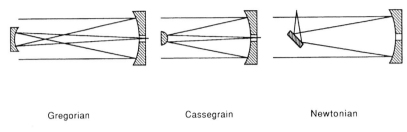

Gregorian   Cassegrain   Newtonian

Figure 1.1. Two-mirror telescope designs.

f number (longer focal length) and thus the asphericalness of his parabola was much easier to grind than the other two models.

At least part of the confusion lay in the poor methods of optical testing in use at the time. Thus, surfaces intended to be parabolic or hyperbolic were probably not accurate. Further, comparison testing was generally performed by the designer and his results transmitted by letter to another observer who compared a verbal description of the view with a competing telescope using different astronomical objects, a different observer at a distant site several weeks later. Judging by the claims and discussions, telescope making in those days was much more an art than a science.[15]

Until optical manufacturing technology could improve, the Newtonian became the instrument of choice among reflecting types, chiefly due to the simpler manufacturing crafts required. While Cassegrain and bent Cassegrain telescopes[16] were produced for special purposes, the Newtonian and its close cousin the Herschellian telescope[17] design were used for almost all large telescopes for about two centuries.

[15] Astronomy is largely a passive science. We observe but we do not touch our subject. There are those who try to simulate or calculate in a laboratory whatever goes on within the core of stars but that's theoretical astronomy. Only in the past two decades have we sent machines and humans to the Moon. We have also sent spacecraft to Venus and Mars to 'taste' the soil and farther to probe more distant planetary systems with radio waves and various sensors. Astronomy as an experimental science is still in its infancy.

[16] The bent Cassegrain looks like the classical Cassegrain but a third mirror on a long rod is inserted up the baffle tube. This mirror sends the light out the declination axle to an instrument or eyepiece. It is popular in an alt-az mounting, allowing a heavy spectrograph to be placed nearer to the instrument's center of gravity rather than hanging it off the back end of the telescope.

[17] A Herschellian telescope is composed of a parabolic primary mirror and an eyepiece situated at the edge of the upper end of the tube. The system thus works as an off-axis reflector. The aberrations induced by off-axis viewing are compensated for by the lack of a secondary mirror which, if it is made of speculum, could absorb about half the starlight.

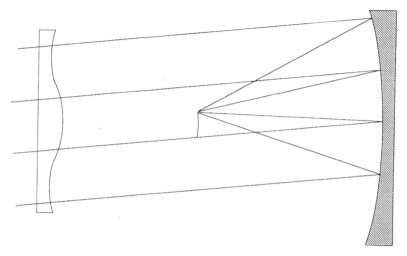

Figure 1.2. Classical Schmidt Camera. Note that the curve on the corrector plate has been greatly exaggerated.

By 1932 Bernhard Schmidt of the Hamburg Observatory was experimenting with manufacturing methods for highly aspheric surfaces. His intent was to produce a photographic telescope – more of a camera – with a wide field of view and a very low f number. The result, as shown in Fig. 1.2, is the classical Schmidt camera.

A thin lens – more of an optically weak correator plate – is placed in front of the mirror. The corrector plate also forms an aperture stop which is typically at the center of curvature of the spherical primary mirror. The corrector eliminates spherical aberrations, an optical defect normally seen in a spherical mirror. The image plane, about half way between the primary and the corrector, still has the defect of curvature of field but the radius is large enough that films could be warped into the proper shape for exposure. Thus, photographs can be made with pinpoint images all across the field.

The trick here is in producing the aspherical corrector plate. In 1936 after Schmidt's death, it was divulged as to how he produced such an unusual curve.[18] He had simply taken a thin optical flat, sealed it to the open end of a cylinder nearly the size of

---

[18] *The History of the Telescope*, Henry C. King, Dover Publications, 1955, p. 357.

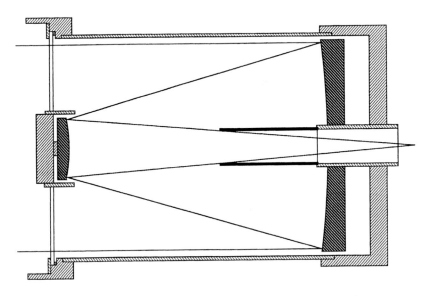

Figure 1.3. Schmidt–Cassegrain telescope cross-section

the corrector and partially pumped the air out of the cylinder, warping the thin plate. He ground the warped surface flat on the exposed side and then released the vacuum, allowing the glass to spring back into the correct shape. While this may seem simple, choosing the cylinder diameter and choosing the correct partial vacuum have remained the 'secret sauce' of Schmidt plate manufacture.

After the construction of several Schmidt cameras, it was realized by a few designers that a spherical mirror placed between the primary mirror and the image plane could redirect the beam through a hole in the primary, thus allowing easier access to the film plane. With the proper radius on the secondary, the image plane can also be made to be flat, making positional measurements of stars much simpler than on a curved film. This is technically a Schmidt–Cassegrain telescope but such designs are usually referred to as cameras rather than telescopes.

The final steps in the evolution of the Schmidt–Cassegrain start with moving the secondary mirror closer to the corrector plate. This results in a smaller secondary with less obscuration of the aperture. Finally, the corrector plate is moved toward the primary which results in a very compact package. One side effect is

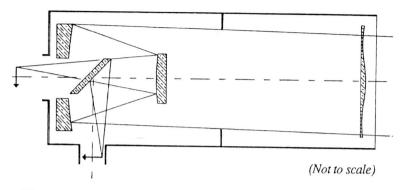

(Not to scale)

Figure 1.4. Bent–Schmidt–Cassegrain showing the placement of an optional tertiary mirror. Illustration courtesy of Coyotè Enterprises.

that the overall focal length of the system becomes a strong function of the radius of curvature of the secondary. Thus, when a secondary of high curvature is used, a relatively long effective focal length can be folded into a short tube. This is especially handy for portable observations or when an observatory would like a larger instrument without constructing a new and larger dome.

While established optical manufacturers would occasionally custom make a single Schmidt–Cassegrain telescope for some special purpose, it was an electronics company that first produced this type in quantity. Tom Johnson and Alan Hale of Valor Electronics had made several telescopes of various sizes and designs starting in about 1954. The first 20-cm f/10 S–C by them appeared in 1966. The reason for settling on that design centered on the portability and ease of use inherent in the type.[19] The company changed its name to Celestronics, then Celestron Pacific and finally Celestron International, reflecting more emphasis on their optical business and less on electronics.

Then in 1971 competition appeared with the introduction of the Dynamax telescope from Criterion Manufacturing Company.[20] Criterion had been making Newtonian telescopes for the amateur market for many years and was expanding its product line. A lawsuit ensued over manufacturing trade secrets, chiefly concerning the method of making the corrector plate.

[19] Private conversation with Alan Hale.
[20] The company later changed its name to Criterion Scientific Instruments.

5

Figure 1.5. Production of 20-cm S–C Telescopes. Each of the 350–500 parts of the instrument is inspected and tested prior to assembly and then the overall system is checked before shipping. Photo courtesy of the Meale Instruments Corporation.

While the suit was eventually settled out of court, ten years of legal briefs piled up before it was over. The Dynamax production facility was later bought by Bausch and Lomb and in 1987 the company stopped making S–C telescopes.

In 1980 Meade Instruments Corporation, long a manufacturer of Newtonian telescopes, started making 20-cm S–C telescopes, providing competition in a lively market. In Japan, JSO and Takahashi have produced S–C telescopes in the 20-cm aperture range.[21] In England, Orion Optics, and in Belgium, Lichtenknecker Optics, have produced the type.

The Lichtenknecker design and one made by Coyotè Enterprises of Des Moines, New Mexico, are interesting in that they are not fixed to one f number or optical configuration. By interchanging corrector plates and secondary mirrors, the tele-

---

[21] The Takahashi instrument actually has a 22.5-cm aperture.

scope can be used either visually or photographically at several different effective focal lengths. This is similar to a system used by Celestron two decades ago to make some of their telescopes more versatile. The corrector plate and secondary mirror were removable as a single assembly. Then a new corrector, tube extender and a film holder were bolted on the front of the telescope, changing it into a classical Schmidt camera.

It is estimated that between 75 000 and 100 000 telescopes of the 20-cm S–C variety have been made to date by the several manufacturers. About 10 000 more are produced each year. While the majority of 20-cm S–C telescopes wind up looking at the night sky, I have seen them used as collimators on optical benches, as sub-millimeter microwave receivers, for artillery spotting purposes, as a boresight camera for a large experimental radar system and for terrestrial viewing as described in Chapter 5. On at least two occasions a 20-cm S–C telescope has gone into space. Once for visual use by the astronauts and once as the optical system for an experiment mounted in the Space Shuttle payload bay. Two of the manufacturers suspect that there have been more uses in space, judging from custom modifications ordered with some telescopes, but the experiments are usually classified. And finally, the type has been used several times as a prop in motion pictures.[22]

---

[22] The only environment I have not heard of a 20-cm S–C working is underwater. It would not surprise me, however, to learn of some aquatic application.

2

First observation – the Moon

Before you set up your telescope you might ask where the Moon is. After a few months of observing you'll have a sense of where the Moon is just as you know, at any time, roughly where the Sun is in your sky. For now, you may have to look it up in one of the popular astronomy magazines. Most newspapers also list the times of Sunrise, Sunset, Moonrise and Moonset in the section with local weather predictions.

Select a site for your telescope which doesn't have trees or streetlamps in the way. After reading your telescope manual, set up the telescope. Now you must align the polar axis to the Pole Star. There are many methods for accomplishing this and your telescope manual probably describes at least one. Another method is described in Appendix 2. Don't let anybody tell you that there is only one 'proper' procedure for polar alignment. The varying procedures differ in accuracy of alignment and time to accomplish.

If you are going to casually visually observe the Moon and bright planets then you don't need to have the polar axis aligned any closer than about three to five degrees. If you are going to look for faint galaxies and nebulae, especially ones which you've never seen before then you need an accuracy of about one degree. This is crucial if you intend to use the setting circles on your mount. It will take three to five minutes to align your telescope this accurately. If you're going to make long-duration photographic exposures of faint galaxies then the error between the polar axis and the true pole must be less than a tenth of a degree. An hour or more may be spent for this alignment. The alignment procedure in Appendix 2 describes the most precise method but it also tells which steps to omit for a quicker, less accurate alignment.

Why observe the Moon first? Well, it's easy to find, easy to focus on and your telescope will reveal a wealth of detail which gives you some understanding of the power of the instrument.

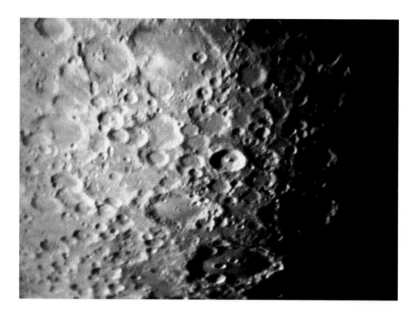

Figure 2.1. Southern highlands of the Moon and the crater Tycho, 1/2-second exposure with tele-extender and 25-mm eyepiece on Kodak #2415 film. Photo courtesy of Meade Instruments Corporation.

Try several eyepieces and see which one works best for you. Most important, observing the Moon provides a basic exercise in getting familiar with the mechanical operation of the telescope. If you should accidentally bump the mount, it's not going to be difficult to reacquire the Moon. You will learn not to lean on the telescope. You will learn how to gently move the field of view from one area of interest to another. You will learn just how sensitive the focus control of your telescope is. You will also discover a new world.

At some point you may come to hate the Moon, for it floods the sky with unwanted light, preventing observations of faint objects. For now, however, enjoy Earth's nearest neighbor. It isn't like most moons, whose diameters are usually 1% or less than the diameters of the planets they circle. The Earth–Moon system, with only a factor of four between their diameters, might more properly be considered a double planet.

Ask most people to envision a Lunar landscape and they won't picture in their minds the classical downward looking 'bird's eye' view of the surface as seen through a telescope.

They'll remember one of the Apollo photographs, for on 20 Jul 1969 the Moon ceased to be an astronomical object and became a real place. Now go back and see it as an astronomical object once more, just as generations of earlier observers discovered it.

You can gaze for hours at the mountains, plains, craters and rills of the Moon. Go ahead and drink your fill at the eyepiece while exploring a whole new world in minute detail. Each night, as the dark shadow of the terminator slowly works its way across the face of the Moon, the shadows lengthen or shorten. Low-contrast shadings appear and disappear. Craters with central peaks loom on the alien landscape. You may even want to compare the view with a good Lunar map in order to learn the names of some prominent features. As a suggestion, find the Sea of Tranquility and see if you can identify the area around Tranquility Base where Apollo 11, the Eagle, landed. You won't be able to resolve the lower section of the Lunar Excursion Module which was left there but you will view a historical site.

While the Moon is best viewed during its partial phases, if you find it near full phase there are a couple of interesting things to look for. First, note the differentiated aspects of Lunar features. There are large, darker areas surrounded by lighter regions with many craters and mountains. Note that the dark areas have significantly fewer craters. The darker features can be seen easily with the naked eye and before the invention of the telescope they were thought to be oceans and seas. Thus, they carry names like Oceanus Procellarum and Mare Imbrium. While we now know that there are no open bodies of water on the Moon, at one time the seas did flow, for they are made up of frozen lava which seeped into the basins, probably after large meteor impacts. These plains are the most recent Lunar surfaces formed and thus they have fewer meteor craters.

When the Moon is full, find the crater Tycho near the South Lunar Pole. There is a system of lighter-colored rays emanating from Tycho stretching across seas and mountains. This is probably material splashed from the crater when the meteor which formed Tycho struck the Moon. Try different eyepieces to see if the contrast between the rays and surrounding terrain is improved. Similarly, around the crater Copernicus there is a smaller system of rays. Look at other large craters under a variety of lighting conditions to see if you can find more ray systems.

Figure 2.2. Southern highlands and Tycho near full phase 1/30-second exposure on Ektachrome 100 film using a 2X tele-extender. Compare with Fig. 2.1 which shows the same region at higher magnification.

Usually, the Sun illuminates only part of the Moon as seen from the Earth. The line separating the bright portion from the dark portion is called the terminator. The dark portion isn't completely black, however. When the Moon is a slim crescent, its night side is illuminated by a nearly full Earth (as seen from the Moon). The pearly image of the dark side can often be seen with the naked eye, especially if the day side of Earth has lots of bright, white clouds. Look at the dark side in your finder telescope and imagine the eerie night as seen from the surface of the Moon with an Earth about two degrees in diameter (four times the size of the Moon as seen from Earth). Now see if you can pick out any features on the dark side. Usually, only the Maria are discernible.

The details of surface roughness are best seen near the terminator for there the shadows are longest. Much study has gone into recording faint changes in shading which would indicate the roughness and texture of the Lunar surface. Just watch a few select areas for several evenings in a row to observe how their aspects change from night to night.

Since the Moon takes about a month to complete its orbit, each 'day' on the Moon is about two Earth weeks long and similarly, each Lunar night is a fortnight also. Thus, the Sun appears

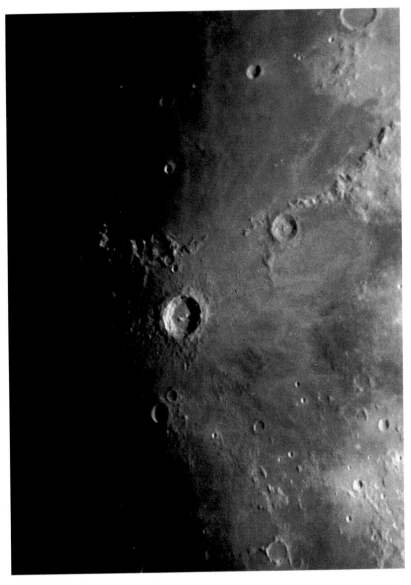

Figure 2.3. Lunar crater Copernicus, 1/2-second exposure with 2X tele-extender and 25-mm eyepiece on Kodak #2415 film. Photo courtesy of Meade Instruments Corporation.

to move much more slowly in the Lunar sky and the Sunrise/Sunset line (the terminator) moves across the surface at only about 15.5 km (9.6 miles) per hour. In some areas the view may be noticeably different after just a couple of hours. It is the night to night variation, however, which will initially show the best observing results.

Some of the larger craters have central peaks and the height of these peaks can be determined by watching their shadows as they walk across the crater floor. The process of determining the height of Lunar mountains is found in several references.[23] The maths gets a little involved but some of the mountains and peaks are many kilometers high. There are two reasons for the larger Lunar mountains. First, gravity on the Moon is only about one sixth of that on Earth so there is less slumping in the rocks. Second, the airless Moon has no weather which would erode the mountains. It is estimated that there are 200 000 craters on the Moon with diameters of a kilometer or more.[24]

Occasionally you might spot a sinuous channel which looks like a river bed. Since there is no water on the Moon how could such a thing be? These are called rills and they are probably collapsed lava tubes left over from when molten rock flowed across the surface. Similar geologic structures can be seen adjacent to volcanoes on Earth. One of the more prominent rills parallels the inner edge of the Appenine Mountains adjacent to Mare Imbrium. Another is located on the flat floor of the Alpine Valley, a remarkable rift which cuts through the Lunar Alps Mountains. This particular rill may have resulted from fissuring and partial filling rather than from lava flow.[25]

Now tour some other interesting places on the Moon. Reading the rest of this book can wait for some cloudy night.

After the observation

Telescope all packed away now? Did you remember to replace the dust cap? Good, how was your first observing session? You

[23] *The Sky: a User's Guide*, David H. Levy, Cambridge University Press, 1991, p. 84. It is also described in *Projects and Demonstrations in Astronomy*, Donald Tattersfield, John Wiley & Sons, 1979, p. 37.
[24] *Principles of Astronomy*, S. P. Wyatt, Allyn And Bacon, Inc., 1977, p. 152.
[25] *Earth, Moon and Planets*, Fred L. Whipple, Harvard Books, 3rd Edition, 1970, p. 116.

may also have learned about wearing warm clothes or buying mosquito repellant. These, too, are an important part of astronomy. You might have found that standing hunched over for long periods has given you a neck or backache. I have a small adjustable-height chair which I use at the eyepiece so you might consider building or buying one, as shown in Fig. 6.3. Most people come away from their first session at the telescope with a list of accessories they're going to buy; a better star chart set, a red flashlight, a card table for all that 'junk' like eyepieces, set-up tools, maps, lists and the ever-present coffee cup. My suggestion is not to buy anything right away. Improvise what you need from around the house, observe for a few more nights and then start acquiring accessories. Get smart on what you need or, like many beginners, you'll buy a huge flashlight more appropriate for signalling ships at sea than reading star charts.

Now, grab that list of things you're going to buy. At the top of it write the date. At the bottom list the objects you observed. This is your observing log. You may want to paste it into a nice notebook or you may, as a friend of mine does, keep scraps of odd-sized papers in shoe boxes. There is a long tradition in astronomy of maintaining an observing log. You don't have to do it but most people get a lot more out of astronomy by making a few notes here and there. Make it as detailed or cryptic as you want. As an example, my observing log for last night is;

17 Jul 91 (18 Jul UT). 20:57 (03:57 UT) arrived Palo Verde Observatory. Observed two total Lunar occultation disappearances for IOTA. Standard optical train & camera, see video data tape #70. Seeing & Transparency excellent. 102 °F. Light, puffy clouds on horizon when I closed the dome at about 22:10 local.

While some observers make copious notes, spending more time looking at the notebook than at the sky, a few straightforward comments will help you recall most of what you observed. You might also want to record things to be done later (buy a new flashlight, look up the Moon's libration some cloudy night) and the weather. Some observers even compose poetry for their observing logs. It's your notebook. Do with it what you will but do give it a try.

How to find your way around the sky by various techniques

There are many methods of finding your way around the sky. I'm not going to tell you which is best because what's best for you may not work for somebody else. It's basically a problem of keeping track of a spherical coordinate system, seen from the inside, and rotating on an axis set at some angle with respect to our conventional coordinates which most of us Earth-bound creatures think of as modified Euclidian plane geometry.

Constellation names and figures have been assigned to various star fields. The groupings are historical and, at times, rather fanciful, being related to ancient tales and fables. These designations are entirely man-made and serve as a convenient method for remembering what stars are where in relation to others. Unfortunately, many of the star patterns do not appear to outline or represent the mythical figures shown on some star charts. Ursa Major, for example, looks to me like a water dipper and its modern appellation of the 'Big Dipper' seems appropriate. Ursa Major, however, means 'Big Bear' in Latin and I don't see the outline of a bear. Cassiopeia looks more like the letter **W** than a lady to me. And Sagittarius looks like a teapot. Maybe it looks like something else to you. Don't feel that you have to memorize the ancient figures. I never did but I do know certain star patterns on sight. They are the patterns which make sense to me. There's a kite with a tail in Aquarius and a scythe in Scorpius. I do, however, see the classical man with a sword in Orion. If you don't like the original constellations, make up your own. Just don't expect anybody else to understand your system, though. We all see different patterns in the stars.

Are the classical constellations useless? No, many people find it easiest to memorize the patterns and knowing them allows you to communicate with other astronomers more clearly. There are also some fascinating tales of Greek mythology woven through their figures. It's good reading for a cloudy night.

Star charts

Star charts are the basic roadmaps of the sky. They are usually drawn in the rotating coordinate system fixed to the stars.[26] Since the stars are seen as if fixed on the inside of a spherical surface and maps are generally flat items, there will be some distortion in the map. That's not a big problem since we've dealt with it before while mapping the spherical Earth (and other worlds).

Maps showing greater areas of the sky have more distortion. For instance, seasonal charts show one half of the entire celestial sphere, as shown in Fig. 2.4. Such maps are usually found in the centerfolds of popular monthly astronomy magazines and are set for about 9 p.m. local time at mid-month for middle Northern (or Southern) latitudes on the Earth. There is extreme distortion at the edge of the map but seldom will we observe objects that low in the sky. The projection scheme of the map shown here retains the shapes of constellations but distorts their overall sizes at the edge of the map. Many people use the map by holding it above their heads, orienting it toward North and comparing with the sky that they see.

If you're going to observe earlier than 9 p.m. local time then use a star chart set for 9 p.m. of the month earlier. If you're going to observe later than 9 p.m. use a chart set for one of the months after the current month. Figure a one-month advance for every two hours after 9 p.m. If you're using magazine centerfold charts then you can't get next month's map until it's published. You can, however, use last year's chart of the same month. The locations of planets and the Moon will be different but the star positions are always the same from year to year. This is also a good reason for saving your old copies of astronomy magazines.

A related map is the planisphere, as shown in Fig. 2.5. This device, made for use over a small range of Earth latitudes, consists of a circular map of the entire sky with the North Celestial Pole at the center (there are Southern Hemisphere versions too). A circular card with an off-center egg-shaped hole is mounted over the map and the two are riveted together so that the card may rotate with respect to the map. Calibrating marks along the

---

[26] The stars aren't actually fixed. Over a period of years they will move slightly and the rotation axis of the Earth will vary. Since these motions are small, a good star chart will remain correct enough for most observations for about 50 years.

August Evening Skies

This chart is drawn for Latitude 40° north, but should be useful to stargazers throughout the continental United States. It represents the sky at the following local daylight times:

Late July	11 p.m.
Early August	10 p.m.
Late August	9 p.m.

This map is applicable one hour either side of the above times. More detailed charts appear monthly in the magazines **Astronomy** and **Sky & Telescope**.

© Abrams Planetarium
Subscription: $6.00 per year, from **Sky Calendar**, Abrams Planetarium, Michigan State University, East Lansing, Michigan 48824.

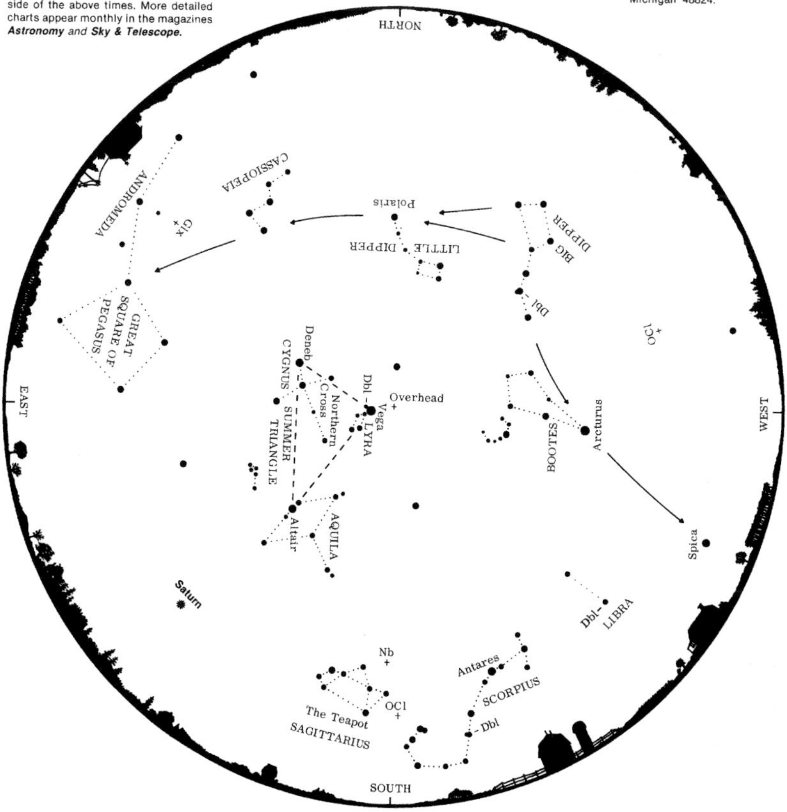

NORTH

CASSIOPEIA

ANDROMEDA

Polaris

M31

LITTLE DIPPER

BIG DIPPER

GREAT SQUARE OF PEGASUS

Dbl

M13

EAST

Deneb

CYGNUS

Northern Cross

SUMMER TRIANGLE

Dbl

Vega

LYRA

Overhead

BOOTES

Arcturus

WEST

Altair

AQUILA

Spica

Saturn

Dbl

LIBRA

Nb

Antares

SCORPIUS

OCl

The Teapot

SAGITTARIUS

Dbl

SOUTH

The planet Saturn is plotted for mid-August, 1992. At chart time 7 objects of first magnitude or brighter are visible. In order of brightness they are: Arcturus, Vega, Saturn, Altair, Antares, Spica, and Deneb. In addition to stars, other objects that should be visible to the unaided eye are labeled on the map. The double star (Dbl) at the bend of the handle of the Big Dipper is easily detected. The double star in Scorpius is somewhat harder. Much more difficult is the double star near Vega in Lyra. The open or galactic cluster (OCl) known as Coma Berenices, "The hair of Berenice," is located between Leo and Bootes. A more compact open cluster is located between Sagittarius and the "tail" of Scorpius. Nearby, "spout" of the "teapot," is the Lagoon Nebula, a cloud of gas and dust out of which stars are forming. Try to observe these objects with unaided eye and binoculars.

—D. David Batch

Figure 2.4. Seasonal sky chart. The maps are produced monthly and have a calendar of astronomical events on the back. Map courtesy of Abrams Planetarium.

Figure 2.5. The planisphere. Illustration courtesy of Sky Publishing Corporation.

edge of the card and the map allow you to line up the local time with the date. When this is done, the hole shows a view of your sky at that time and date. The edge distortion is a bit greater than the all-sky seasonal charts and it becomes extreme for lower latitudes. It is, however, a very handy device and I keep one at my desk for planning observations. Planispheres and other star charts may be purchased via advertisements in the popular astronomy magazines, at telescope dealers or at map shops.

Seasonal charts and planispheres show only the brightest stars. Fainter and more interesting objects such as nebulae require more detailed maps. One of the handier formats is the 'orange peel' map which is simply a flattened spherical map section or gore, usually covering about five hours of right ascension

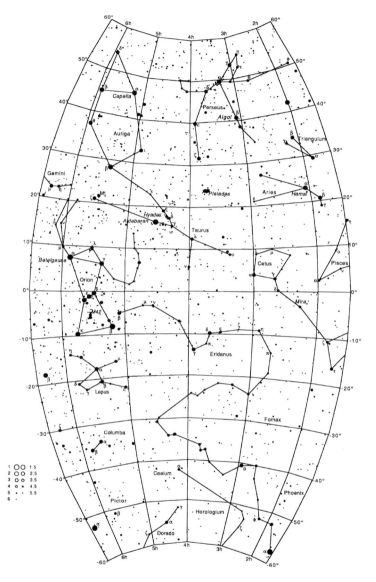

Figure 2.6. Orange peel star chart.

from about 60° North declination to 60° South declination, as shown in Fig. 2.6. Two circular maps covering the poles complete the set. My favorite version of this type of map is *Norton's Star Atlas*. I use it at the eyepiece and have worn out about four editions of the book in my observing life. Indeed, it has been said that you can tell how experienced an astronomer is by seeing how many bugs he has squashed in his Norton's. The book also has an excellent reference section with everything from definitions of astronomical terms to hints on how to ease stiff eyepiece draw-tubes. Each of the eight large charts has a list on the back of interesting double stars, variables, nebulae and galaxies. Stars are shown down to about Mv 6 and all of the Messier objects are plotted. The bounds of the Milky Way are shown along with many faint nebulae.

I have made a template for use with Norton's charts which has holes of several sizes in it. The holes correspond to the fields of view of both my finder scopes and some of the lower-power eyepieces when used in the main telescope. Thus, I can see at a glance that to reach a particular object all I might need to do is put the finder on a bright star, then move two finder field diameters to the right and up a half a field to view. This technique is known as star hopping. More about that later.

An alternative to Norton's is the *Skalnate Pleso Atlas of the Heavens*. This is a collection of 16 unbound charts which come in either a desk edition (white background with black stars) or a field edition (black background with white stars). The chart projection is more like a standard Mercator map. I find the 46-cm (18-inch) by 30-cm (12-inch) format to be a bit too large to handle in a breeze so I've photocopied the charts with a 2:1 reduction into notebook-sized pages. The print is a bit small but that doesn't bother me. It does bother several of my observing companions so try the reduction on one sheet before you spend the money on reproducing the whole set. See if you can read the print when your eye is dark adapted.[27] I have had the reduced set of charts laminated in plastic so that the night dew doesn't make the pages soggy.

There are more detailed star charts available but they are very expensive and typically not of much use to the beginner. *Norton's Star Atlas* will keep most people busy for a year or two. When the time comes to search out FFN[28] objects then I'd recommend either the *Smithsonian Astrophysical Observatory Atlas or Tirion's*

---

[27] The dark-adapted eye is wide open and often can't focus as well on near objects as it can in broad daylight. This is especially so for people who are borderline farsighted.

[28] Faint Fuzzy Nothings.

Sky Atlas 2000.0. Typically, Tirion and the SAO are large, heavy books and thus are too unwieldy for use at the telescope although I have at times photocopied a page from these atlases and taken that to the eyepiece.

How deep in magnitude do you want to go? If your limit is about Mv 9 then you will have to contend with an atlas containing about a quarter million stars.[29] The accompanying four volume catalogue listing star data such as positions, spectral types and proper motions is much too heavy to haul on field trips. Since I received a copy of the SAO Catalogue on computer disk I seldom use the hardback volumes because it's easier for the computer to search for information. I haven't given the volumes away though, because they're still useful for elevating visiting minors at the dinner table.

For deeper surveys there are photographic atlases. These are matched sets of precision photographs of the sky, taken at standard positions and overlapping adjacent photos on each edge. Clear plastic overlays allow you to determine the right ascension and declination of any star. Hans Vehrenberg's *Falkauer Atlas* is one of the better ones and comes in two versions, one with stars to Mv 14 and the other to about Mv 17. The smaller of the two with a limiting magnitude of 14 consists of 428 unbound photographs. Each is about 18-cm (7 inches) square and covers a 10° square portion of the sky plus almost a degree of overlap with adjacent plates. The larger version, showing even fainter stars uses prints almost twice as big. Finally, there is the *Palomar Optical Sky Survey* (POSS). Professional observatories use these huge contact prints, taken with a 1.22-m (48-inch) aperture Schmidt camera. They show stars to about Mv 22. I wouldn't recommend buying this set of prints just now since the total cost is about five times the price of a new 20-cm S–C telescope. Be aware, however, that the set is available at most good observatories and universities and you can check out any puzzling object with them. If you can see it in your 20-cm S–C then it will be easily visible on the POSS.

A celestial sphere is a model of the heavens painted on a globe. Often they are made of clear plastic so that models of the Sun, Moon and some planets can be positioned within. I own one and

---

[29] The SAO Catalogue has 258 997 entries.

Figure 2.7. A star dome. Illustration courtesy of Sky Publishing Corporation.

it looks horribly scientific, lending an aura of scholarly dignity to the den but I seldom use it. Many people do, however, and some take it to the telescope. The problem I find is that you look at a celestial sphere from the outside and you look at the sky from the inside. Some people can look at a sphere, perform the transform in their heads and find objects in the sky.

There is one ingenious version of a celestial sphere which is useable at the telescope. Imagine painting the stars of a celestial sphere on the outside of a rubber ball. Now put a metal ring inside the ball to stiffen the sphere at its equator (but let the ring slide around freely). Now suck all the air out of the ball and while you're doing this, keep one half of the ball as round as you can while pushing in the other side until it fits neatly inside the rounded half and forms a concave hemisphere. The metal ring will keep the open end of the hemisphere circular. If the inside of the ball is slippery enough you can slide the surface of the ball around and place any desired portion of the painted surface inside the concave volume. The device shows stars only to about fourth magnitude and thus it is about as useable as the seasonal star charts or a small planisphere. It does, however, accurately represent the sky as a three-dimensional inverted bowl overhead. For those people who just can't make the transition from a distorted two dimensional seasonal chart to the real sky, this device may help.

Computers seem to be doing everything these days from cooking our food to dynamically tuning our automobiles' engines. The desktop computer has placed powerful calculating capabilities in just about anybody's hands. There are many inexpensive computer programs which will plot star charts for observation planning. Some astronomers even drag the computer out to the telescope and use it not only as an electronic atlas but as a data recording medium, an electronic scratchpad for observing notes. While I wouldn't recommend buying a computer just for its ability to generate star charts, if you already have one then this is an ideal use for it.

One of the problems inherent in making star charts is that stellar data bases take up large chunks of electronic storage space. The SAO Catalogue in abbreviated format (RA, Dec and Mv only) requires about 10 million bytes of data. That data base covers all stars down to about Mv 9 but no non-stellar objects such as galaxies and nebulae are given. As memory systems become cheaper, this problem will fade. Similarly, as processors and co-processors become faster, the time to plot charts and print them will decrease. The first star plotting program I worked with about eight years ago took upwards of 20 minutes to plot a 5°×5° area in the sky. That system also lacked large mass storage devices so about every two or three minutes I had to feed the computer a new floppy disk of star data. Today's computers can keep the whole data base on a hard drive and plot the same area in under a minute.

I use computer generated star charts to observe asteroids. I can plot a wide field (10° × 10°) picture down to, say Mv 6 and then make a more magnified map of the area of interest. The charts shown in Fig. 5.1 were made this way. The path of the asteroid can then be added along with tick marks at the beginning of each day. Usually I have the computer precess the coordinates[30] to the

---

[30] This is a calculation which accounts for the slow wobble of the Earth's axis and is commonly referred to as the precession of the Equinox.

epoch date of the observation. Then I'll ask the computer if there are any interesting nebulae, galaxies, comets or other asteroids nearby which might confuse me. While I don't have a complete list of asteroids, the computer stores the brighter ones on disk. A little work with the all-sky display which resembles the seasonal charts lets me know if there will be a problem with a nearby bright Moon or twilight conditions. Planet locations are added into the display just to see if there will be anything interesting nearby. You never know when you'll have a few extra minutes at the eyepiece.[31]

This capability came in handy when I heard about a predicted occultation by an asteroid I'd been trying to observe for months. The faint asteroid would pass in front of a slightly brighter star, blocking its light for several seconds and allowing a measurement of the diameter of the asteroid. I'd been foiled on four previous occasions in measuring this asteroid; clouded out, equipment problems, out of town on a business trip and (I am ashamed to say) I slept through the fourth chance, having forgotten to set my alarm clock. I looked forward to nailing down that asteroid's diameter at last until I went to the computer and found that the Mv 10 star would be four degrees from a nearly Full Moon at an elevation of only 13° above the horizon well into dawn twilight. Sleeping in that morning was a wise decision. My telescope would have been washed in the early morning downpour which occurred during the time of the observation. On the other hand, I now know that in about a year and a half, another opportunity will present itself with that same asteroid high in the sky occulting a much brighter star in the dark of the Moon during Arizona's best season of weather.

The computer is probably best used as a planning tool, generating finder charts for difficult-to-locate objects and allowing the observer to avoid observations under adverse circumstances. As such, it is an accessory to the telescope but it's not often used at the eyepiece. There are those, however, who attach the computer to the telescope to either control its motion or take data. Such capabilities will be discussed in the sections about electronic setting circles, photometry and TV imaging.

---

[31] Star chart generation and planetarium software exists which can also control the pointing of automated telescopes.

The cognitive map; best star chart of all

We all carry in our heads a map of where we are on the surface of the Earth and how to get to our workplaces, shopping markets, etc. There's a less well defined region outside of this area showing places that we visit only intermittently. Beyond that we either need a map or we get lost. The same is true in the sky. Novice observers will be able to find the Pole Star (Polaris) and only a few constellations. As you become more experienced, you will come to remember, for instance, that the Ring Nebula (M57) is between the two Southern stars in the parallelogram South of Vega (Alpha Lyra), as shown in Fig. 4.4. If I look at that piece of sky, I imagine I can almost see the Ring where I know it to be. Intellectually, I know I can't possibly see the faint Mv 9 planetary nebula with my naked eye but I have viewed it so often that I don't need a map to find it with my telescope.

While this may seem like some exotic mental trick, it's a common ability among astronomers. One talent which I do not possess, but many observers do, is a knowledge of what constellations are above the horizon at any given time. While I've remembered a few key constellations at various seasons, I'm often forty or more degrees off in predicting the altitude and azimuth of a given celestial object at a particular time if I try to do the problem in my head. I have a friend, however, of whom I can ask, 'Where is Arcturus (Alpha Boötis) now?' He will point immediately in the correct direction to within about five degrees, even if he's sitting inside a building. He will point to some spot on the floor if the star has set. He also seems to know the local and sidereal times to within a couple of minutes without consulting his watch. His is a talent I envy but then again we each have a talent. Why did mine turn out to be the ability to remember exactly the text of soap commercials heard thirty years ago?

Star hopping

Star hopping is a combination of the cognitive map and a star chart approach. It is the commonest method of finding your way around the sky. Even though I have accurate setting circles on my telescope I usually star hop because that's the way I first learned to find things in the sky and it seems most natural to me. You may find another method which feels comfortable. While I've operated large, computer controlled telescopes in which the observer sits two floors below the dome at a console and merely types coordinates onto a computer terminal and then commands, 'MOVE', finding a particular faint object in the sky manually does give a small sense of accomplishment. In star hopping you start with your cognitive map and think of the star you know nearest the object which you wish to find. Then you drag out the star charts and identify the star you knew already.

From that star you then identify nearby patterns of fainter stars, working your way across the star chart until you get to the object of interest. At the eyepiece, you first slew to the known star manually and set the clock drive going. If there will be a significant North or South slew from the known star to the object of interest, check the declination tangent arm to assure that there is enough travel to reach the object. You might also check to make sure the finder scope is aligned to the main optics. To do this, place the widest field of view eyepiece you have on the main telescope. Place the known star in the center of the finder. Does it fall generally near the middle of what you see through the main optics? If not, center the star in the middle of the main telescope and then adjust the screws on the side of the finder so that the star is centered in the finder telescope.[32]

When you look through the finder telescope you probably won't see the star field oriented as you'd expect it. Most straight-through finders have a double reversal of direction. Stars expected at the top of the image are at the bottom and stars expected at the left side are on the right. It's not difficult to correlate the finder image with the star charts, however. Simply rotate the star chart 180° as you hold it in your hands so that the top is now at the bottom and the left side of the chart is now on the right. That should make the picture match reality.

Right-angle finder telescopes, however, have an extra reversal of direction in one dimension only, usually top to bottom. You

---

[32] In some of the cheaper telescopes, the finder scope flexes with respect to the main optics. Thus, if you align the finder with the main optics while observing an object at the Western horizon and then slew to the Eastern horizon, the finder and the main optics will no longer be exactly aligned. I once had this problem when I bought a cheap large aperture finder. I've since cured it by rebuilding the finder mount but I still check it often.

have five choices; you can try to perform the image flip in your head to match what you see with the charts. Sometimes I can do this but often I get confused, especially if I'm tired. The second alternative is to flip the star chart face down and try to shine a flashlight through it from the front side, thus reading it backwards. This sprays bright light all over the observing site, a socially unacceptable act at polite star parties. Third, you can do as one owner of a 20-cm S–C has done and have all of your charts photocopied backwards at considerable expense. Fourth, you can replace the elbow (diagonal mirror or star diagonal) in the finder with one which has a special prism rather than a simple flat mirror. This compensates for the extra reversal of direction. Fifth, you can replace the right-angle finder with a straight-through finder which has an image that is easier to interpret. Of course, the straight-through finder will cause you to assume some odd contortions and positions, crane your neck until it aches and generally make you wish you'd kept the other finder. My solution is that I have one straight-through finder and one right-angle finder on the telescope. I use the straight-through finder until I have to line up on something at high elevations, causing me to half-kneel (and there's always a sharp pebble where I put my knee), crane my neck and wonder why I didn't take up stamp collecting. Then I switch to the right-angle finder and live with the image reversal problems. Life is full of compromises.

Once you have identified the known star in the finder and oriented yourself to the star charts, you can now compare the view with star patterns in the direction of the object of interest. Slowly move the field of view until your original star is at the edge. Check the new stars which have appeared at the other side of the finder field of view. Are they the ones shown on the star chart just a little farther from the known star? If not, go back and check your orientation.

Now continue moving the field of view slowly. Pause every third of a field of view or so to check what you see against the charts. Once you let that guidepost star slip out of the eyepiece, you're in unexplored territory. If you become unsure, retrace your steps and move back to the one familiar star which you started with. It may take several backtracking passes before you venture farther toward your objective.

If the move involves both right ascension and declination, it may be easier to move first in one axis and then in the other rather than to try a diagonal move. Deciding which axis to move first should depend on where the best star patterns are. Pick a route through the star fields with the most distinctive set of star patterns. With a little thought you can see lines of stars, parallelograms and pairs of double stars.

After a few months of returning to favorite objects you will have memorized the routes from known stars to faint but rewarding scenes. Once you find that you don't have to consult the charts to travel the well-remembered route then your internal cognitive map will have been expanded. I've found that some star-hopping routes which I learned as a child and haven't used for a decade are still fresh in my memory. I suppose it's like going back to your old elementary school for a visit. You still know where the water fountain and the principal's office are.

Setting circles

Almost all 20-cm S–C telescopes come with setting circles. Unfortunately, some have 7- or 8-cm (3-inch) diameter markings and some are cast into the declination axis end plate. In general, the larger the setting circle diameter then the more accurately can readings be determined. One S–C telescope manufacturer apologized for the poor resolution of his setting circles by explaining that the circles were intended only to be accurate enough to put any desired object somewhere within the finder telescope (which has a field of view of about 5°). I consider that the circles should be able to point to within ±1° in order to be useful. The 10-cm (4-inch) diameter setting circles on my C-8 are able to accomplish this but only if I am very careful about polar alignment and calibrating a start point on the indicators. About half the time I can place the desired object within the 1° field of view of the main optics when using my lowest power eyepiece.

In order to use the setting circles you must first align the polar axis, as described in the owner's manual or in Appendix 2. You will need at least one degree accuracy. Next, slew the telescope to some bright star with a known position in the sky. Appendix 8 lists common bright stars. Note that the star list should have an epoch or date within twenty-five years of the current date. Otherwise, the stars will have drifted slightly with the precession of the Equinox and the setting circle will be in error. It is easiest and most accurate if the star is near the meridian and near the Celestial Equator. Center the star in the main optics. On the telescope base is a setting circle calibrated in hours. There is a pointer on the rotating part of the base on the South side (North side for Southern Hemisphere observers). Move the setting circle until the correct hour mark is aligned with the pointer. The circle may be a bit stiff to turn, especially in a new telescope, but it should move with respect to both the base and the rotating part which supports the fork arms.

Once you have aligned the hour circle it should have enough friction that it will move with the rotating part of the base as the clock motor drives the telescope. When you unclamp the right ascension axle to slew the telescope, the hour circle should remain moving slowly with the drive clock. Most hour circles will remain accurate for several hours. If you are not using a precision drive corrector, however, the clock motor which drives the telescope will be about one degree off after a six-hour observing session. If you turn the drive motor off to go inside and warm up with a cup of coffee then the right ascension setting circle will have to be recalibrated.

Now check the reading indicated on the declination setting circle against the published declination of the star. You may have to loosen the central screw to adjust the declination circle. Some S–C telescopes have a setting circle on each side of the declination axis. Be sure to set each one. Once you have performed this setting, I wouldn't expect to have to adjust the declination again unless the telescope has been boxed and shipped with rough handling. Still, I check both circles each time I observe. Once, after six years of use, the screw holding one of the circles came loose and I had to recalibrate it.

Now you are ready do dial a star. This is perhaps a faster method of finding objects but you'll miss discovering many interesting objects chanced upon when star hopping.

About time

Astronomy is one of those sciences where knowing the time is essential.[33] Indeed, our clocks are set by astronomical observations. While different observers around the world operate on

---

[33] It has been quipped that time is Nature's way of assuring that everything doesn't happen all at once.

various local clocks, all astronomers use Universal Time (UT), formerly called Greenwich Mean Time and sometimes called Zulu Time. This is the local time at Greenwich at 0.0° longitude. It does not shift back and forth by an hour at various seasons as Daylight Saving Time does. The main reason for using one time for all observers is so that events which can be seen by several observers in different time zones can be related to one master clock.[34] The current UT may be calculated by adding or subtracting the appropriate number of hours from a local clock (noting that a date change may also be in order) or it may be determined from radio time signals broadcast by various agencies.[35] I have a very handy little radio which is pre-set to the frequencies of WWV, a station run by the US government and the radio also receives weather reports. Most national timekeeping agencies also have automatic phone answering machines which give accurate time. I would caution against banks and local radio stations which give the time as a free phone-in service. They are often inaccurate by a minute or two.

A second time which is useful to astronomers is Mean Local Time, not to be confused with Local Standard Time. Local time zones are set approximately 15° of longitude wide and the time in that zone is Mean Local Solar Time (Local Standard Time) for the central longitude of that zone. An astronomer, however, may live at the edge of the time zone and thus the Sun rises and sets up to a half hour earlier or later than at the central longitude of that time zone. In order to calculate the Mean Local Time for your observatory, subtract four minutes from the Local Standard Time for each degree of longitude that you are West of the central meridian of your time zone (or add four minutes for each degree East).[36] This becomes crucial when determining the times of Sunrise, Sunset, Moonrise and Moonset – events which can seriously hamper observations.

[34] For readers in North America, the Central Time Zone during standard time seasons (non-Daylight Saving Time) is 6 hours earlier than UT. I leave it as an exercise for the student to figure out how to determine UT in England.

[35] In North America, observers can receive radio station WWV or WWVH at frequencies of 2.5, 5, 10, 15 and 20 MHz or radio station CHU at 3.330, 7.335 and 14.670 MHz. In the British Isles the frequencies are 5, 10 and 15 MHz. In addition, BBC Radio 4 at 198 kHz gives accurate time checks. Finally, non-verbal coded time data are broadcast at 60 kHz in both North America and the British Isles.

[36] As an example, I live at about 112° West longitude. The central longitude for my time zone is 105°. Thus, I subtract 28 minutes from the Local Standard Time to obtain Mean Local Time.

Finally, many astronomers use Sidereal Time which is equal to the right ascension passing your local meridian at any moment.

A word on maintenance and cleaning

In general, preventive maintenance is better than required maintenance. Try to keep the telescope out of the dust and rain. While some manufacturers' instructions indicate that the telescope should be placed in its box between observing sessions I don't do that since it's too much trouble to re-erect the tripod, wedge, etc. And some day I know I'm going to drop the tube assembly if I lift it out if the box enough times. I leave the telescope and mount assembled and wheel it into the garage each night. Then I throw a plastic trash bag over it and secure that with an elastic cord.

Your telescope will accumulate dust, scratches and wear marks through the years. For the non-optical surfaces the dust can be whisked off with a dry brush or cleaned with a moist rag. The scratches and wear marks are badges of honor, indicating many hours of use. They don't detract from optical performance and besides, when I observe the lights are off so I can't see them.

Resist the temptation to clean your optics. When your eyes are dark adapted and you shine a flashlight at the corrector plate, every minute speck of dust looks like a rock. It is permissible to use a little 'canned air' to blow loose dust off of the corrector and I do this before each observing session.[37]

More corrector plates have been harmed than helped by cleaning them. This is because the rubbing action of cleaning can grind small dust particles into the glass surface and may produce permanent scratches. In addition, many cleaning compounds leave a residue which blocks more light than the dirt they removed. Although a corrector plate may look terrible in the oblique light of a flashlight, realize that most glasses transmit only 95% to 98% of the light. The 2% to 5% reflected can look terrible to a dark-adapted eye but even if you cleaned the corrector perfectly you'd still see that reflected light. It's not dirt – it's the normal surface of the glass. The decision to clean your corrector

---

[37] If you have access to a fixed installation air compressor, beware of using it until you check to see if oil is added to the air stream. Air compressors used for driving power tools often have an automatic oil injection device. This helps lubricate the power tools but it also produces a messy oil film on your optics.

plate is about as serious as the decision to rebuild your automobile's engine and it should probably be done just as often. A description of how to clean the corrector plate is given in Appendix 4.

After every road trip or shipment of the telescope by air, I make a quick check to assure that the bolts and screws holding the telescope together are going to continue performing that function. Other than a focus knob set screw losing its grip and a couple of missing finder telescope alignment screws, I haven't had many problems. There have been reports, however, of the declination clamp and the declination slow motion controls of some models working loose. These are finicky little mechanical devices and the fact that their design has been repeatedly changed shows that the manufacturers are working on the problem.[38] I have checked the manuals of the three major manufacturers and all describe how to (repeatedly) adjust these components. I'd agree that this is a design defect in the telescope but if it's your worst problem in astronomy then your problems are not very great.

After shipping the telescope, it may require collimation. The procedure in Appendix 3 describes how to do this. If you have never done it before then it is best to have someone who is experienced show you how to do it. If you get confused about which adjustment screw to turn which way (and there is no standard among manufacturers) then you can wind up with worse collimation than when you started the procedure. When in doubt consult the telescope owner's manual or the telescope dealer.

As telescopes and other complex mechanical devices go, the 20-cm S–C is remarkably maintenance free. Let me emphasize once more that the best maintenance program is to prevent the telescope from becoming dirty or banged up in the first place. Repairs are always more expensive than prevention.

---

[38] One observer, a mechanical engineer by trade, designed a foolproof declination clamp requiring no adjustment and replaced the original model with his own. Sensing that he could make a profit from his invention, he investigated the possibility of a limited production run and found that each unit would cost $110.00 for machining alone. After assembly, shipping and advertizing were factored in, it would have cost $200.00 per telescope.

3

Planets, double stars and other bright things

Before you view a planet you must first find one. For many people this is a non-trivial problem. You can approach it as the ancient people did and watch every night for objects which slowly move amongst the stars. Five planets – Mercury, Venus, Mars, Jupiter and Saturn – were discovered this way. There is another clue and that involves the twinkle of the stars. The stars are so distant that they can be considered, for all practical purposes, point sources of light. As their light comes down through the atmosphere, small variations in density, temperature and upper atmosphere winds act like weak lenses, focusing and defocusing the light from moment to moment. Thus, the brightness of stars appears to change. Their position in the sky will also appear to change rapidly, causing the star to appear as a small fuzzy disk in the telescope. This is the atmospheric seeing limit discussed earlier.

Planets, on the other hand, cannot be considered point sources of light, for they have a measurable, if small, disk image. The light from one side of the planet does not travel the same path through the atmosphere that light from the other side travels. Each atmospheric disturbance thus affects only a small part of the light coming from the planet. The effects of many separate tiny disturbances tend to cancel each other out when looking at a planet with the naked eye. Thus, planets do not appear to twinkle as the stars do. It should be noted that on nights of really terrible seeing, you can detect twinkling in some planets if you look for it. This is rare, however. On such nights the viewing isn't going to be very good. The only two observations you can make on evenings like this are naked eye meteor observations and admiring the Aurora.

Simply knowing when a planet will be above the horizon is half the battle in finding it. Planetary rise/set times are charted in

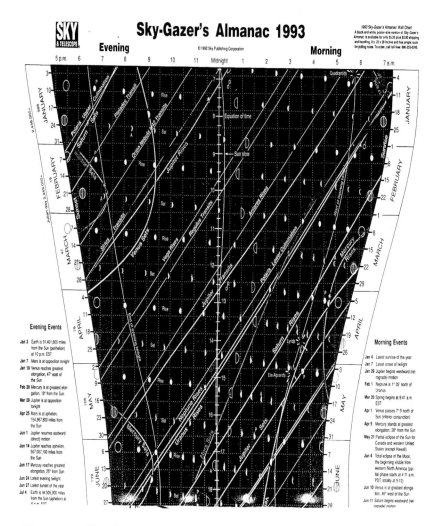

Figure 3.1. Sky-Gazer's Almanac. © 1992 Sky Publishing Corporation, reproduced with permission.

the Sky-Gazer's Almanac which is usually the centerfold of the January issue in popular astronomy magazines. This is a two-dimensional graph in which the vertical axis is marked off in the months, weeks and days of the year. The horizontal axis shows the time of night from dusk to dawn. A planet's slowly varying rise/set time can be plotted as a curved line on this chart. Similarly, the chart indicates the beginning and end of twilight, the rise/set times for the Moon and a host of transient events such as eclipses. It is one of the more useful planning tools I have.[39]

Simply knowing when a planet will rise doesn't necessarily allow you to find it, especially if the planet is on the other side of the Solar system and thus faint. Planets can be found by referencing the popular monthly astronomy magazines and their seasonal sky charts. Excellent computer programs also exist which make finder charts for planets.

Once you have finally located the planet and captured it in the eyepiece, what next? Mars, Jupiter and, to a lesser extent, Saturn, all show constantly varying weather patterns. They rotate and, within a few hours, will show a completely different view. From night to night the patterns will change. The outer gas giant planets have several moons which can be seen with a 20-cm S–C although you may have to watch for an hour or two to differentiate between real moons and faint background stars.

Often the planets look best at dawn or dusk. This is a subjective opinion and is hotly debated among visual planetary observers. One school of thought is that with a brighter sky background, the veiling glare of the planet's bright image is lessened and thus more detail can be seen. Another possible reason is that during twilight your eyes aren't fully dark adapted. Thus your pupils are smaller and the outer parts of your eyeball lens (where the greater aberrations reside) are covered. In addition, in some locations the best atmospheric seeing is recorded just before Sunrise. This is a very localized phenomenon but I have lived in an area where this was true and found that it was well worth getting up early. Each observer is encouraged to explore this phenomenon at a local site.

Mercury and Venus, the inner planets

The inferior planets (so called because they have smaller orbits than our own) of Venus and Mercury will show phases like the

---

[39] When using the Sky-Gazer's Almanac, always be sure to include the correction for Mean Local Time as described in Chapter 2.

Moon.[40] They will usually be in either dawn or dusk twilight.[41,42] I have often observed these planets if, after I set up my telescope, I note that the sky is too bright for more serious work and I must wait for the end of astronomical twilight. Similarly, after a night of viewing, dawn often arrives too early but I want just one more pleasant observation before I toddle off to bed. Venus and Mercury are good targets under these circumstances.

After you have observed that Venus often shows a crescent phase like the Moon what else is left to see? Venus has no natural moons and our exploration spacecraft orbiting the planet are far too faint to be seen with even the largest Earth-based telescope. Using various colored eyepiece filters, visual observers have often reported seeing faint cloud patterns on the Sunlit side. The Association of Lunar and Planetary Observers (ALPO) coordinates efforts at documenting these changes. Their address is given in Appendix 1. Similarly, markings on the dark side, often called the 'ashen light', have been reported to ALPO. While these observations are controversial and require visual discrimination of minute shadings, they may, in the end, be useful. Many long hours at the eyepiece are required just to begin to understand the subtleties of Venus's image. But then again, that doesn't daunt the observer who is looking for just such a project which will keep him at the eyepiece often and for long periods.

Mercury is a tough object for beginning planetary observers since it never appears farther than 28° from the Sun. Thus, it is usually found low in the sky and its image is distorted by the atmosphere. It is, however, interesting to watch it swim in waves of seeing, for occasionally the atmosphere will steady down for a moment, allowing a glimpse of its crescent shaped sunlit side. Since Mercury is observed low in the sky, its

[40] On November 6, 1993, an interesting observation was possible for the planet Mercury when it passed between the Sun and the Earth, appearing as a black dot on the Sun's surface. The transit of Mercury will be visible to observers located in Hawaii and Australia. Venus will Transit the Sun in 2004 and again in 2012.

[41] Venus has often been called Earth's sister-planet, for with a diameter of 12 300 km it is 81% as large as the Earth. It also has a rocky core and an atmosphere but that is where the simlarity ends. Its thick, poisonous atmosphere blocks out all but a dim red glow from the Sun and temperatures at the surface are high enough to melt lead. Its environment has often been likened to that of Hades.

[42] An interesting article by David J. Eicher describing the observation of Venus appears in *ASTRONOMY Magazine*, November, 1991, p. 89.

Figure 3.2. Venus 1/2-second exposure on Ektachrome 400 film using eyepiece projection (25-mm ocular) to increase magnification. Photo by Mark Coco, courtesy of Celestron International.

image is often dimmed and reddened by passing through the atmosphere.

Mars

About once every two years the Earth, in its much faster orbit, catches up with and passes Mars. For a few months, the Red Planet looms high and bright in the night sky.[43] With medium to high powers[44], the planet reveals a wealth of visual detail, from polar ice caps to curious markings which change from week to week.

As our nearest planetary neighbor in space, Mars has been the subject of intense study, dating from the time of the first crude telescopes. The tantalizing sight of a whole world blurred by the limit of atmospheric seeing has caused several observers

[43] Because the orbit of Mars is slightly elongated, some encounters, or favorable oppositions place us closer to the planet than others. At its very closest the planet can come within 56 million kilometers (35 million miles) of Earth. On some oppositions the planet will come within only about 97 million kilometers (60 million miles).

[44] At a favorable opposition, Mars can attain an apparent diameter of 25 arc seconds. Thus, with an eyepiece of only 75 power the planet will appear as large as the Moon does to the naked eye.

Figure 3.3. Mars 1-second exposure on Ektachrome 400 film using eye-piece projection. Photo by Steve Edberg, courtesy of Celestron International.

to imagine slightly more detail than actually exists on the Martian surface. While several spacecraft have sent back high-resolution images of surface features, useful information can still be obtained by visual observers studying weather patterns on Mars. Huge globe-circling dust storms occasionally arise in its thin atmosphere, obscuring most of the surface. After one of these episodes, some darker markings on the planet appear to have changed shape. This is probably caused by deposition of lighter-colored dust over darker-colored features.

In the course of a Martian year, the sizes and shapes of the polar ice caps can be seen to change, for Mars has Summer and Winter seasons alternating in the Northern and Southern Hemispheres just as on Earth. There are other similarities between the two planets. For instance, Mars rotates in the same direction as Earth with a period of 24 hours 37 minutes. This implies that if you observe Mars every night for a week, you will keep seeing pretty much the same hemisphere each evening. After a few weeks, though, you will be able to study Mars' other side.

There are several features of Mars which you will not be able to see easily. The two tiny moons, Phobos and Diemos, are so

Figure 3.4. Mars, 3-second exposure by Raul Espinosa on Kodachrome-64 using a 9-mm eyepiece projection system and a 2X tele-extender.

faint that they are lost in the glare of the planet for all but the largest telescopes. Similarly, the volcanoes and craters of Mars are usually too small and too low in contrast to observe except at the best of observing sites on very rare nights of good atmospheric seeing. Even when they are spotted, their identity is known only through confirmation by spacecraft images, for before those images were made, Mars observers could only speculate on the probable nature of the markings.

In spite of seeing problems, Mars does paint an interesting picture of larger changing features. As Summer progresses in each hemisphere, the polar ice caps shrink and adjacent areas darken. For years, many observers thought this a sign that vegetation sprouted, watered by the melting ice. Alas, on Mars there are no plants as we know them and the darkening effect has been linked to seasonal wind patterns moving light-colored particles of dust from one place to another.

In the early days of Martian observations, astronomers sketched the image by hand, referring to the eyepiece often. For long minutes, the observer might sit motionless at the eyepiece, recording nothing but waiting for that moment of perfect seeing when the Earth's atmosphere would suddenly calm down and reveal a steady image. After drinking in as much detail as possible, the observer would hastily but carefully draw every feature and subtle shading from memory, taking care not to embellish the image with preconveived notions. Analysis would come later but at the telescope the observer was to be simply a recorder. Indeed, the job title of recorder is still used in some astronomical organizations for people employed in this and similar jobs.

For years photography was employed to overcome the limitations of the eye and record impartial images. Unfortunately, exposures of up to a second are required on most telescopes and the atmosphere moves too much in that time. Thus, ground-based photographs of Mars usually have less resolution than sketches. Recently, CCD camera images of Mars have been processed in computers to compensate for the effects of atmospheric seeing. This technique shows great promise.

In spite of technology, I still enjoy the visual impression of Mars and can spend quite a bit of time exploring this beautiful world. I know of several amateur astronomers who have sketched and photographed the planet for years, compiling a private study of each favorable opposition. And when Mars has disappeared behind the Sun for a few months, they can still review notes and sketches, enjoying hours of astronomical relaxation in the company of their favorite planet.

Jupiter

Locate Jupiter and look at it in your finder scope. Don't use the main optics yet. You will see a small ball and three or four of the brighter moons lined up. Do you think you could learn much from this image? Well, the finder scope on most 20-cm S–C telescopes closely approximates the magnifying power of Galileo Galilei's original six to eight power telescope. Study the view carefully and extract as much information as you can in order to understand Galileo's problems with the limits of the state of the art in his time. By the way, the glass transparency, eye relief, field of view and optical resolution of your finder scope is much better than Galileo's.[45]

---

[45] Many years ago several replicas of Galileo's telescope were made using authentic materials and fabrication methods. I have observed with the one at the Flandrau Planetarium in Tucson, Arizona. It's not much fun to use. It's not as good an optical instrument as a toy plastic telescope I found as a prize in a box of cereal.

Figure 3.5. Jupiter, 4-second exposure on Tech Pan film with an effective f number of 138. Photo by Rouse, courtesy of Celestron International.

Look for the four satellites. They should be about Mv 6–7. Occasionally you may see only three or two. Perhaps one of the satellites is behind the planet or in its shadow. Look closely at the ball of Jupiter. You may just barely be able to detect a sharp dark spot (verify this in the main optics) caused by the shadow of one of the inner satellites passing between Jupiter and the Sun. With keen eyesight you may also see the satellite causing the shadow in front of the planet. It is difficult to spot the moons of Jupiter when they are in front of the planet because the moons resemble some of the smaller weather features on the planet. At very high magnifications on nights of excellent seeing, these moons do show a disk and thus can be differentiated from background stars.

You may find it interesting to draw the Jupiter system each night for a week or so and determine for yourself which of the satellites is orbiting closer to the planet and which are farther. The four brightest Jovian satellites are called the Galilean satellites after their discoverer. They were named Io, Europa,

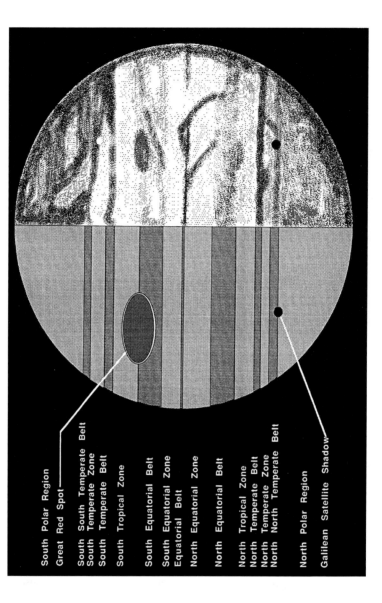

South Polar Region

Great Red Spot

South South Temperate Belt
South Temperate Zone
South Temperate Belt

South Tropical Zone

South Equatorial Belt

South Equatorial Zone
Equatorial Belt

North Equatorial Zone

North Equatorial Belt

North Tropical Zone
North Temperate Belt
North Temperate Zone
North North Temperate Belt

North Polar Region

Galilean Satellite Shadow

Figure 3.6. Jupiter's visual details. The classical belt designations are given on the left but as can be seen in the sketch on the right, the planet doesn't always conform to classical theory. Note the dark circles which are shadows of the Galilean satellites.

Ganymede and Callisto, from innermost to outermost, by the German astronomer Marius who independently discovered them only one day after Galileo first saw them on January 7, 1610.[46] Jupiter has a dozen or more fainter satellites.[47] It takes a bit of observing to differentiate field stars from Galilean satellites. This is a task which Galileo himself had to accomplish in order to show that the satellites do indeed circle Jupiter, a controversy at the time.[48] Thus, Galileo's simple observation shattered long-cherished misconceptions with a small array of facts which anybody with a telescope could verify. Having demolished the Earth-centered model, it may seem a simple matter to place the Sun at the center of the Solar system, a theory several learned folk had advanced over the ages. Unfortunately, Galileo owned one of the only working astronomical telescopes available and the local Establishment proved extremely reluctant to accept new ideas. One hopes that modern-day observers will have better results in showing their peers how the Jupiter system works.

Now look at Jupiter in the main optics. Pump up the magnification using an eyepiece of 150X or so. Later you'll try higher powers if the seeing is steady enough. Now you should see at least two dark bands or belts on the planet. As you watch, subtle gradations denote more smaller belts. The planet becomes more interesting the longer you look at it. The Great Red Spot could be on your side of the planet too. Then smaller white spots may become visible. This is a complex place, not suitable for just a quick look. Better drag the observing chair over and get comfortable first.

Jupiter has a system of rings surrounding the planet inside the orbits of the brighter satellites. The rings, discovered by passing Voyager spacecraft, are not visible from Earth except via sophisticated, large telescopes and image processing. The rings and satellites all rotate in about the same plane – much like a miniature Ecliptic. That plane is tipped with respect to Jupiter's orbital

---

[46] *Astronomy*, Robert H. Baker, Van Nostrand Company, 1964, p. 222.

[47] Every time somebody sends a spacecraft past the planet a couple more are discovered and frankly, I've lost track of the number of football field sized rocks orbiting Jupiter. The numbers increase and the pieces get smaller until they form a very faint, hazy ring of dust particles (visible from spacecraft) circling the planet.

[48] At the time of first observation in 1610, current scientific wisdom had it that everything revolved around the Earth and nothing revolved around any other object.

plane about the Sun by only 3°. This means that we see the Jupiter system nearly edge-on and its satellites quite often can pass in front of each other or cast shadows on each other. The satellites also disappear into Jupiter's shadow occasionally. This makes for quite a dynamic scene which some astronomers measure and time with high precision.

Now go back and look at the weather patterns on the planet. Jupiter is a gas giant planet which means that it is mostly methane, ammonia and hydrogen with a small rocky core. Of course, we see only the cloud tops while most of the planet's secrets are hidden. Jupiter displays huge cyclonic storms such as the Great Red Spot, which is so large that you could fit the planet Earth inside it easily. This rotating storm has been watched for over 300 years. The spot is most easily seen in detail when it transits the Central Meridian (the longitude on Jupiter facing Earth) or is at least near the Meridian. Since Jupiter completes a rotation in less than 10 hours, you won't have to wait long to see the Spot.[49] Indeed, you can see the entire planet in one night. Having seen it all, however, doesn't mean that you are finished with the planet, for during 10 hours the weather changes considerably. Small white spots have formed and dissipated. The bands may change their width or coloring. The Great Red Spot may have moved slightly or changed color. Indeed, for several years the spot faded and became difficult to see. Recently it has become more visible but it appears to me less red and more yellow. Many observers use colored eyepiece filters to enhance the contrast of these features. Jupiter watching and recording has become the sole observation of some astronomers who nightly sketch or photograph the planet.

Jupiter, the king of planets, has one interesting aspect; it actually radiates heat energy in the infrared. This is the energy released as it contracted from the primordial cloud of gas and dust that formed our Solar system. At the center of Jupiter the temperature may be considerable. It is not sufficient, however, to fuse the hydrogen in its atmosphere which would make it a small star. About ten Jupiters are required to generate sufficient gravity to heat up gases to fusion temperatures. Thus Jupiter is, in a sense, a failed star.

---

[49] Jupiter is so large and rotates so rapidly that it bulges appreciably at its Equator.

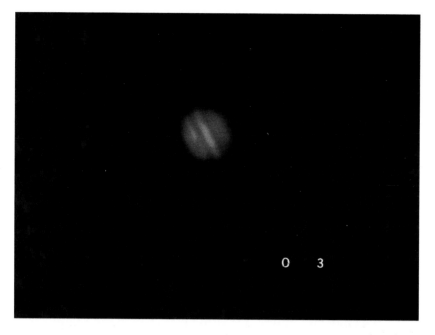

Figure 3.7. Jupiter, 3-second exposure by Raul Espinosa on Kodachrome-64 using a 9-mm eyepiece projection system and a 2X tele-extender.

Saturn

Your first impression of Saturn will include the beautiful ring which surrounds the planet.[50,51] A second glance may reveal a small, dark gap near the outer edge of the ring. This is Cassini's Division and it is best seen at the East and West edges of the rings.The rings are made up of billions of particles of ice and

[50] When I was just a lad learning the planets, one of the burning questions in astronomy asked why Saturn and no other planet had rings. Eventually we learned from spacecraft that Jupiter, Uranus and Neptune have faint rings which are difficult to see from Earth. Now one of the great questions in astronomy asks why planets like Earth don't have rings. For a fascinating discussion of the discoveries of rings, see *Rings*, James Elliot and Richard Kerr, The MIT Press, 1984.

[51] Saturn revolves around the Sun with a period of about 29 1/2 years and the rings are inclined 27° to the plane of the planet's orbit. Twice during each Saturn 'year' the rings are nearly edge-on as seen from Earth and they seem to disappear for several weeks. If your first look at Saturn is disappointing in that the rings aren't visible, then you have the rare opportunity to looks for Saturn's inner moons which are faint and often discernible only when the glare from the rings is absent. The rings will reappear in a few months. The next occasion when the rings will be seen edge-on is in 1994–95.

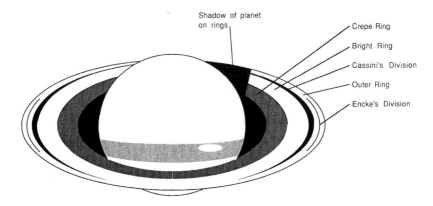

Figure 3.8. Typical features in the Saturn system.

dust, each speck orbiting independently around Saturn. Thus, while the rings look solid, they are actually a cloud of debris.

Generally, the rings are classified as the Outer Ring (outside Cassini's Division), the Middle Ring (also called the Bright Ring) and the Crepe Ring or Inner Ring. This last ring is faint and its separation from the Bright Ring isn't easily discernible.[52] On very clear nights of steady seeing a second gap, known as Encke's Division can be seen in the middle of the Outer Ring. This is a very difficult feature to discern and my observing logs indicate that I have seen it only three times in forty years – and one of the occasions required a 1.5-meter aperture telescope designed specifically for planetary viewing. The other two times involved my trusty 20-cm S–C so the observation can be made. It's probably not so much a matter of observing experience but more a problem of finding a site with steady seeing.

For years amateur and professional observers intermittently reported seeing many divisions within the rings and spokes or radial features which rotated with the rings. I looked often and saw none. Photographs didn't record the features and they were dismissed as optical illusions, much like the canals on Mars. Dedicated observers reporting such things were scoffed at. Then

[52] The great planetary observer Gerard P. Kuiper often commented that Cassini's Division was the only real division in the rings and that there is no gap between the Bright Ring and the Crepe Ring. See *Astronomy*, seventh edition, 1964, by Robert H. Baker, D. Van Nostrand Company, Inc., p. 225.

Figure 3.9. Saturn, 4-second exposure by Raul Espinosa on Kodachrome-64 using a 9-mm eyepiece projection system and 2X tele-extender.

spacecraft with high-resolution cameras flew near the planet and found thousands of tiny rings, resembling the grooves on an old phonograph record. They also found spokes. These are not well understood phenomena and the Association of Lunar and Planetary Observers (ALPO) welcomes reports of observations of them. Spokes are low-contrast phenomena requiring excellent seeing and atmospheric transparency.

One of the more interesting and dynamic observations that can be made with the rings is when a star passes behind them and is dimmed by the ice and dust. Such events happen only once or so per decade but they are well worth seeing. On July 3, 1989, I watched for several hours as the star 28 Sagittarii passed behind the rings, winking on and off by the second as it was occulted by individual chunks of the rings and occasionally peeking dimly through dusty areas. The data were recorded on video, providing a permanent record of the density of the rings. Since I had little to do during the observation other than feed fresh tapes to the video recorder, I was free to sit back and enjoy the sight as the star shone brightly through Cassini's Division and then abruptly disappeared at its edge, reappearing faintly seconds later. With a gang of astronomical friends nearby, we marvelled at the dynamics of the observation. It was quite a memorable evening.

Often observers concentrate so much on the rings that they ignore the planet itself. Saturn exhibits belts, spots and other weather features just as Jupiter does. The only problem is that due to a haze in Saturn's upper atmosphere, the image appears washed out and doesn't show much contrast. Some observers have tried eyepiece filters of various colors in order to enhance the visibility of the markings.

Depending on the relative positions of the Earth, Sun and Saturn, one can often see the shadow of Saturn falling on the rings and the shadow of the rings falling on Saturn. This lends a three-dimensional aspect to the image of the Saturn system and allows an appreciation of just how huge the rings are. With a diameter of about 275,000 km (175,000 miles) you could just about slip an Earth-sized object between the inner edge of the Crepe Ring and the planet.

Like Jupiter, Saturn is a gas giant planet and it has collected many small moons. While spacecraft have catalogued a score of chunks circling the planet, only nine moons will be visible in your 20-cm S–C telescope. They are Mimas, Enceladus, Tethys, Dione, Rhea, Titan, Hyperion, Iapetus and Phoebe. They will appear like stars near Saturn and the only clue that they are actually moons will be that they follow the planet and not the sky. Titan is interesting in that, although a moon, it is larger than either Pluto or Mercury. It also has an atmosphere with reasonable air pressure at the surface but the air composition consists of nitrogen, methane and hydrogen cyanide at freezing temperatures. While it is not a hospitable place to Earth creatures, it may harbor life. Contemplate that while you gaze at its image.

Iapetus is interesting in that its apparent brightness changes dramatically, varying between Mv 9.5 and Mv 11.0 as it circles the planet. Since the moon is tidally locked to Saturn (like our own Moon, its rotation period equals its revolution period and thus one side always faces Saturn) this means that one side is dark and the other side is light colored. Phoebe exhibits an unusual orbit in that it circles in the opposite direction to all the other bright satellites. This probably indicates that it was originally an asteroid captured by the gravity of Saturn and was not formed at the same time as Saturn and its other moons.

Uranus

Uranus was discovered by William Herschel in 1781 as he performed a systematic sweep of the skies with his home-made telescope. It is curious that nobody had ever recognized the planet earlier since at a maximum brightness of Mv 5.5 it is just barely visible with the unaided eye at a dark site. On several occasions I have seen it without a telescope but then again, I used a star chart to find it and knew what I was looking for. Still, in more than a century and a half between the invention of the telescope and the discovery of the planet, over 20 observers had recorded Uranus and marked it down in their logs as a star.[53,54]

It was not the anomalous motion of the planet among the stars that made Herschel notice it, but the fact that it shows a perceptible, if small disk of 3.5 to 4 arc seconds in diameter. Thus, it does not appear point-like as the stars do. Perhaps the discovery had to wait until telescope technology attained sufficient resolution to reliably resolve such small disks.

When observing Uranus it is best to focus the telescope on one of the nearby moons or field stars which are indeed point sources. Then glance back to the planet and you will see the small green disk. Herschel at first mistook Uranus for a comet because of its non-stellar appearance. Many observers have reported bright spots and bands on the planet but with such a small target, excellent skies and considerable patience are required to make the observation. Some Uranus buffs have tried with moderate success to see more detail by inserting blue or orange eyepiece filters in the optical train.

---

[53] *Astronomy*, seventh edition, 1964, by Robert H. Baker, D. Van Nostrand Company, Inc., p. 230.

[54] Young and impressionable students of astronomy are often treated at this point with the story of 'Sloppy Pierre', an observer who did not keep a neat and orderly observing log and thus missed discovering a planet. In researching the tale, however, I found one reference that the story might be apocryphal. I found several astronomers who remember hearing the story but, like myself, could find no reference to it in their class notes.

As with other planets, some visual astronomers maintain that twilight conditions are the best time to observe because the glare of the planet is not so great against a brighter sky background. The reasoning is that with a little light from the background, the pupil of the eye closes down slightly, covering the outer portion of the eye's lens which is more likely to have aberrations which defocus the image.

Uranus has two moons barely visible by eye in a 20-cm S–C telescope and three more which can be photographed easily. Titania at Mv 14 and Oberon at Mv 14.2 are brighter while Miranda at Mv 16.5, Ariel at Mv 14.4 and Umbriel at Mv 15.3 are fainter. Like the Jupiter and Saturn systems, these moons look like a miniature Solar system with one difference. The planet's rotation and the plane of the moons' orbits is tipped almost 90° with respect to the orbital plane of the planet about the Sun. Thus twice each 84 year journey around the Sun, it points either its North or South pole at the inner Solar system and we see its moons describing nearly perfect circles about the planet. Uranus has several smaller moons which are not visible in a 20-cm telescope.

Uranus has a faint set of rings just as all the other gas giant planets possess. They are not visible from Earth but they can be detected even in small telescopes when a bright star passes behind them and is momentarily blocked by the dark ring material. Indeed, this is the method by which they were initially discovered when their existence was not even suspected.[55]

Neptune

Neptune was discovered with pencil and paper and not by telescope. Urbain Jean Joseph Leverrier and John Couch Adams independently worked at the problem of Uranus's position not matching its theoretically calculated orbit. They postulated in the early 1800s that yet another planet must be gravitationally influencing Uranus and set out to calculate where that planet might be. Both astronomers reached solutions pointing to the same area of the sky at about the same time in 1845–6 but they had troubles

---

[55] For a fascinating discussion of the discoveries of rings, see *Rings*, James Elliot and Richard Kerr, The MIT Press, 1984.

locating a telescope to make the observation (they could have used a 20-cm aperture telescope and found it easily). Leverrier won the race to find a telescope and Neptune was finally observed on September 23, 1846, by Johann G. Galle.[56]

This blue-green planet reflects the weak light of the Sun from the outer Solar system and shines at about Mv 8. With a revolution period of 165 years, it creeps very slowly across the constellations so it will take several nights to confirm that Neptune has moved. With careful observation you may be able to discern its 3.2 arc second disk. While some observers have reported markings such as bands, they were not seen on high-resolution images from passing spacecraft.

According to my logs, I have not observed Neptune very much. When I do, it is usually on one of those nights which occurs once every few years in which all the planets are visible at some time in the night. On those occasions a few friends gather and we do the 'Grand Tour' and at each stop compare notes and tell tales about observing the planet of interest.

One observation I have made often is watching a star pass near the planet in search of occultations by rings. For years, we knew that Saturn, Uranus and Jupiter all had ring systems but we had no evidence of any rings around Neptune. Any star coming even remotely close to the planet was observed dilligently by members of the International Occultation Timing Association (IOTA) for any dimming which might indicate the presence of material between the star and the telescope. It would appear from spacecraft reconnaissance that Neptune has ring arcs or rings with uneven density around their circumference. Alas, it was my luck to observe either through the thinnest part of the ring or the thickest clouds of the year, for I never caught a ring occultation. I shall keep trying, however.

---

[56] A comedy of errors and missed communications combined with the clash of egos of some renowned astronomers of the day served to delay the first observation of Neptune. The tale is told in *Astronomy*, Fred Hoyle, Crescent Books, 1962, p. 164, *Earth, Moon and Planets*, Fred L. Whipple, Harvard Books, 3rd Edition, 1970, p. 34, and *The Sky: a user's guide*, David H. Levy, Cambridge University Press, 1991, p. 132.

Figure 3.10. Pluto, showing the planet's motion over a period of 48 hours; 30-minute exposures at prime focus with Kodak #2415 film. Photo courtesy of Meade Instruments Corporation.

Pluto

At a maximum brightness of Mv 13.7, Pluto is a tough planet to see visually in a 20-cm telescope, although it is easily photographed. Finder charts are usually given in the January issues of most popular astronomy magazines. It is a difficult object because, at 2300 km in diameter, it is small and it is also the farthest planet.[57] It's moon, Charon, can't be discerned as a separate object because of the limits of seeing. Charon, at 1200 km in diameter, is only about half the size of Pluto and the pair has thus been dubbed a double planet rather than a planet with accompanying moon. Give this tiny object a try, for there is a certain satisfaction in knowing that you have seen all of the planets in the Solar system.

Double stars

It turns out that we on Earth circle a rather odd star in that it has no companion star. Most stellar systems are made up of pairs and even triplets of stars but our Solar system has only one. This is probably a good thing since two local stars would probably make finding a clear, dark night for observing much more difficult.

Double stars come in two varieties; the first, and less interesting type, is star pairs which appear close together in the eyepiece but which are not actually involved with each other. Indeed, one may be twice as far from us as the other. They merely lie roughly along the same line of sight. The second type represents stars which are gravitationally bound to each other and which rotate about each other.[58] These are of interest to astronomers, for by knowing the parameters of the orbits the stars make about a common center of gravity, we can calculate the mass of the system. Usually the stars take a few years to circle each other so the measurement is a leisurely one.

---

[57] At the time of writing, Pluto is actually slightly closer to the Sun than Neptune. As it travels its elliptical orbit of 248 year period, however, it spends the majority of its time outside the orbit of Neptune.

[58] Binaries are further classified as either visual or spectroscopic binaries. In visual binaries both stars can be clearly seen while for spectroscopic binaries the separation is too small to be discerned in a telescope. Spectroscopic binaries are discovered by seeing the spectra of two stars superimposed on each other, often with Doppler shifts indicating their co-orbital velocities.

Alpha Centauri requires 80.1 years to orbit its companion.[59] They, in turn, are circled by a distant, faint companion, Proxima Centauri, making the system a triple.

Three- and four-star systems are not uncommon. Indeed, it has been estimated that at least 5% of all visual binaries have one or more companions.[60] A typical example is Epsilon Lyræ, shown in the finder chart for the Ring Nebula in Fig. 4.4. At first glance, this is a widely separated double star with the components about 3.5 arc minutes apart at visual magnitudes 4.5 and 4.7. A closer inspection, however, reveals that both of these stars are binaries with separations of just under three arc seconds, and thus the system is a double-double. For Southern observers, the star Beta Tucanæ presents a similar double-double view. Finally, the star Alpha Capricorni is a system of at least six stars. The primary stars are separated by about six arc minutes. Both are visual doubles. In the fainter one, the secondary pair is separated by about seven arc seconds and the fainter of those two is itself a double with a separation of 1.2 arc seconds. At least one other star in the system is a spectroscopic binary.

The components of a binary need not be similar, as they are in Epsilon Lyræ. Odd pairings of widely divergent star types occur often. One common example is Albireo (Beta Cygni).[61] It is a yellow and green pair (some observers see yellow and blue) with strikingly different colors. The different colors in stars are a partial indicator to their age (along with mass and luminosity). Thus studying stars of different colors often helps astronomers determine the life cycle of different types of stars.[62]

---

[59] If the reader is unfamiliar with the nomenclature of stars, Appendix 7 attempts to sort this out, realizing that the system of naming stars, nebulae and galaxies is haphazard at best.

[60] *Astronomy*, seventh edition, 1964, by Robert H. Baker, D. Van Nostrand Company, Inc., p. 414.

[61] Albireo is not a true double star since the components are quite far from each other in distance but nearly along the same line of sight.

[62] Because stars age so slowly, altering their appearance over periods measured in tens of thousands of years, astronomy has often been called a static science. We can see both old and young stars but relating the two is, at times, a bit of a chore. It is as if somebody handed you a photograph of a forest showing not only mature trees but also seedlings and rotting logs lying on the forest floor among dead leaves. Could you, from that single photograph, describe the life cycle of a tree? Could you describe their relationships to the squirrels, the grasses and the bacteria in the soil? Now relate all that to the paper in the photograph.

Double stars are often used to test the resolution of telescopes as the observer tries to 'split' faint, close pairs. While this is a fun contest and often settles bar bets among telescope owners over whose instrument is better, it is more often a test of the seeing and how well different apertures handle atmospheric interference.

Another optical test is to look at double stars with widely varying magnitudes. While the fainter star may have been visible if it were alone in the field, the light and glare from the brighter one may wash the dimmer companion out. This is a test of unwanted reflections from the corrector plate and it is essential that the optics be clean to minimize scattered light. One of the toughest doubles of this class is the star Sirius (Alpha Canis Majoris), which is the brightest star in the sky. It has a companion of Mv 8.6 which orbits with a period of about 50 years. The separation between the two stars varies from three to eleven arc seconds.[63] Unfortunately, the companion is approaching the primary star and they will be closest in about the year 2000. Better wait until 2025 to try this one.

For those who are interested in long-term projects, plotting the separation and position angle[64] of binary stars is an interesting hobby. A quick check of some double star tables, however, reveals that most doubles observable with a 20-cm S–C telescope have periods in excess of 50 years and some are estimated to be in the thousands of years. Still, after a decade or so, some motion is discernible in the shorter period examples.

Optimizing the view

First, get comfortable – you can see more that way. There are those astronomers who maintain that you're not doing real astronomy unless you're in an uncomfortable position freezing your star charts off. Don't believe them.

---

[63] I know several people who claim to have split Sirius with a 20-cm S–C telescope. I have been trying for the past year and a half to perform the feat but have not been successful. Perhaps when my eyes were younger and my youthful enthusiasm more intense, I might have seen it. I shall keep trying, however.

[64] The position angle of the fainter component is measured with respect to the brighter one. The angle is measured from a line connecting the primary to the North Celestial Pole to a line connecting the primary and the secondary, reckoned in degrees from North around through the East.

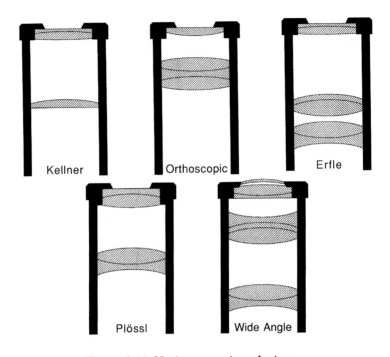

Figure 3.11. Various eyepiece designs.

Selecting the optimum eyepiece is rather like picking up the correct utensil at a formal dinner table equipped with ten or more pieces of silverware per place setting. For any given dish, perhaps half the possible tools will get you food but one will do the job best. There are four primary measures to eyepieces; magnification, eye relief, apparent field and cost. Magnification is simply the ratio of the focal length of your telescope (2000 mm) to the focal length of the eyepiece. While higher magnifications may initially seem desirable, excessive magnification dims the perceived surface brightness of faint objects and, in the extreme, simply magnifies atmospheric seeing problems. Eye relief is the distance from the eyepiece to your eyeball while seeing the entire apparent field of view. For most people, a spacing of a few millimeters is comfortable. In some eyepiece designs such as the Kellner when used at high power, the eye relief is uncomfortably short and this can easily induce fatigue in the eye. The apparent field is the solid angle over which the observer sees an image when he looks into the eyepiece. Given two different eyepiece designs at the same magnifying power, one may show a larger section of sky than the other and this is the apparent field of view. Cost is, of course, a factor since some observers may invest a third of the cost of their telescope in eyepieces, eyepiece filters and Barlow lenses alone.

The Kellner design is a simple, inexpensive eyepiece yielding good images at lower powers. At high magnifications it suffers from low eye relief and a small apparent field of view. Orthoscopic eyepieces offer better eye relief and a wider apparent field while adding color correction, an important factor in planetary observations. The Plössl design is sharper at the edge than the Orthoscopic with good contrast and color correction for Lunar and planetary observations. Some astronomers also use them for deep-sky work. The Erfle eyepiece has a very wide apparent field and good eye relief when used at low to medium power. Some of the less expensive examples, however, suffer from lack of sharp focus at the edge of the field. Erfles are best used for deep sky observations.

Recently, several wide-angle specialty eyepiece designs have come on the market. Used mainly for deep-sky work, they have apparent fields so large that one often cannot see the whole view without moving the head from side to side. With seven or eight elements, some examples suffer from light loss due to absorption but modern anti-reflection coatings appear to be helping the problem. They are very expensive but the views afforded are often worth the cost.

Two other eyepiece designs not shown in Fig. 3.11 are the Ramsden and Huygenian. These are older designs with fewer elements and less performance. They are generally not found on newer telescopes.

While a wide selection of eyepieces is certainly desirable, the expense is high. Thus, an invention called the Barlow lens is often included in the optic accessory box. This is a small negative lens in a tube which slips into the eyepiece holder. Its other end is designed to accept a standard eyepiece. The Barlow lens effectively doubles or triples the magnifying power of the eyepiece at the expense of image brightness. While this may seem an inexpensive method of effectively doubling the number of eyepieces in your collection, the images often lack clarity and are not as sharp as those seen through an eyepiece of half the focal length without the Barlow lens.

Eyepieces are generally 31.75 mm (1.25 inches) in diameter but there are two other standard sizes; 24.5 mm (0.965 inches) and 50.8 mm (2.0 inches). Adapters are available for the other sizes and generally the 24.5-mm eyepieces will work well with a 20-cm S–C telescope. They are less expensive and the only drawback concerning them is that in the longer focal lengths (lower power) they do not have as wide a field of view as the larger eyepieces. The larger-diameter eyepieces work well in lower power applications such as rich-field viewing, often giving spectacular images of large chunks of the sky. If, however, you are using a 90° 'star diagonal' mirror to bend the light path upward to a more convenient angle before it enters the eyepiece, realize that a star diagonal designed for 31.75-mm (1.25-inch) eyepieces will cut off the light at the edge of the field when used with a 50.8-mm (2.0-inch) diameter eyepiece. Most major telescope manufacturers make a larger star diagonal and eyepiece holder as an optional accessory.

I have tried several zoom eyepieces in various telescopes. The eyepiece has a knob which, when turned, changes the effective focal length of the eyepiece. I own one of these devices with an effective focal length of from 8.4 mm to 21 mm. While this might seem handy, I find that fixed focal length eyepieces yield a better image at all settings. Perhaps this device shows its strength in terrestrial nature studies where one might have to track some rapid activity in the eyepiece and can't afford the few seconds to change eyepieces. Other than showing other astronomers the device as a curiosity and occasionally demonstrating its poor performance, I have never had it out of the eyepiece box.

My favorite eyepiece is a 20-mm Erfle. It has good eye relief, a wide field of view (65°) and reasonable sharpness at the edge of the field. I once found a good 5-mm Orthoscopic for an excellent price at an astronomical swap-meet. In the ten years which I have owned the eyepiece, I have pulled it out of the accessory box only three times. Each time, I have looked at the 400X image and decided that all I was magnifying was the seeing. Some night I may encounter that moment of perfect seeing when I can crank up the magnification as high as I want and I'll be glad I bought that eyepiece. Until then, I haven't gotten my money's worth out of it.

The eye as a detector

The human eye is both a wonderful detector of photons and a fairly poor scientific instrument. Only within the last few years have we developed electronic devices which are as sensitive in real time (without hours of integration) as the eye. Yet, the eye is inseparably connected to the human brain which processes visual information. While the processing is vital to understanding images, it can be misleading. The eye/brain combination can be, and is often, fooled by what the brain thinks it sees and this leads us to the realm of optical illusions.

A common example of this is off-axis vision. Look at any star field. A dense field such as may be found in the Milky Way is best. Find the very faintest star you can see. You might find that by looking directly at the star it seems to disappear yet while looking slightly off to one side the star pops back into view. The reason for this disappearing trick is that within the eye on the image-forming surface there is a small area of lesser or no sensitivity where the optic nerve enters the eyeball. Visual observers soon learn to scan across the field of view and pay attention to the area adjacent to this 'dead zone'. This is commonly referred to as using averted vision. Why don't we see this phenomenon in broad daylight? The dead zones for the left and right eyes don't fall on the same place. When working in broad daylight with both eyes, the brain fills in the missing pieces in each image with data from the other eye and thus you 'see' photons which fell in the dead zone.

A second example is seeing faint detail, say on a Lunar or planetary image. The atmospheric seeing may make the image swim in the eyepiece. For fleeting instants the image appears to settle down and then break up once again. The brain records those moments of seemingly best seeing and believes those images to be true. The problem is that the image processing can be too complete. Four or more tiny transient dark spots on the Lunar surface may be interpreted as a line connecting the dots. The brain's experience in poor seeing or lighting conditions tells it that this is probably the case. Unfortunately, the brain is no longer dealing with familiar terrestrial views and the image may indeed be four unconnected dots.

Giovanni Schiaparelli observed Mars visually during the favorable 1877 opposition and thought he saw many straight lines which he called canali. This was an unfortunate choice of words, for canali means either channels or lines in Italian. People started calling them canals in English and a flurry of observations resulted with wild speculations about intelligent life on Mars. Alas, the Mariner and Viking spacecraft set the record straight that there are no canals on Mars. It has often been quipped that intelligent life may be involved in Martian studies but the real question is which side of the eyepiece the intelligence lies on. Even with the sure and firm knowledge of the Viking images in my mind, I once observed Mars with the Lowell Observatory 24-inch (61-cm) Clark Refractor on a particularly calm night and saw the lines. My own image processing brain told me there were objects before me that I knew did not exist.[65]

While the eye is an imperfect observing device, it is probably the very best sensor most of us will ever own. There are, in addition, some things which we can do to enhance the view. The first and most obvious is to maintain dark adaptation. We are all familiar with the phenomenon of walking from a well-lighted building into the night. At first we can see nothing but within seconds images begin to form using night vision. After a minute or so the dramatic increase in sensitivity slows but it will still develop even more with time. Night vision uses a completely different set of optical nerves than day vision. These more-sensitive photon detectors use a chemical which is destroyed by bright light and it takes the body many minutes to manufacture the chemical in sufficient quantity. Indeed, the common wisdom[66] is that 45 minutes of dark adaptation are required before full night vision is achieved. I do know that the briefest flash of white light will set the process of night vision adaptation back almost to the starting point. Thus, it is considered extremely poor form to spray large quantities of photons at a dark-adapted astronomer.[67]

Red light does not seem to destroy night vision adaptation as much as other colors. Thus, when lighting is needed, as in read-

---

[65] Lest the reader think that I experienced some special privilege using Percival Lowell's private telescope, you might be interested to know that on at least one night a week the telescope is open for public use. Call Lowell Observatory at (602) 774-3358 in Flagstaff, Arizona to obtain a schedule.

[66] Common wisdom = most astronomers know this fact but can't remember where they heard it. Furthermore, it seems to agree with their own experience

[67] The usual comment from astronomers who have thus been abused is, 'If I could find the idiot who shined that light at me I'd strangle him.'

ing star charts or adjusting equipment, a red light will allow you to retain dark adaptation. I'd recommend using the minimum intensity of light since very bright red lights will inhibit some night vision.

Most evidence suggests that night vision is color blind. Thus, the pretty colors of nebulae seen on long-exposure astrophotos will not appear visually in your eyepiece. Most nebulae appear gray visually. While some stars do show color, they are so bright that they approach day-vision light levels across a pointlike area. Only on the planets will color become an interesting view. These, like bright stars, have light levels when seen in a 20-cm telescope that are comparable with a well-lit room. I have noticed often that after observing bright planets or the Moon I need to wait many minutes before my complete night vision returns.

Much has been written on the avoidance or ingestion of substances and certain activities in order to enhance night vision. My own experience is that it depends on the individual. I know that one glass of wine cuts my own acuity by about one visual magnitude yet I observe with a friend who appears unaffected. The same is true of tobacco and most non-prescription drugs such as cold remedies. Altitude (and lack of oxygen) decreases almost everybody's night vision. High altitudes will also make the observer more susceptible to errors in logic and common sense, almost as if alcohol were involved. More than one astronomer has gone to a high peak and made long time exposures using an empty camera. Professional astronomers on mountain-top observatories often use bottled oxygen while observing. I know of one amateur from a sea-level city who took a very expensive vacation high in the Rocky Mountains to do astrophotography. The local observers who helped him fared well but he ruined every shot, reinforcing the lore that only long term (years) of acclimatization will overcome severe altitude problems.

From my own experience I know that fatigue inhibits visual acuity and have heard that eating a large meal just before observing is not good, although I've violated this rule many times without encountering disaster. I have several friends who gulp down various vitamins and I've tried it but there doesn't seem to be much difference in my case.[68] It would appear that different people react to various substances and situations differently. You must therefore study your own observing habits and rate your visual acuity and sensitivity. If, some night, you don't seem to be seeing all of the stars in the Pleiades cluster (M45) that you normally do, reflect a bit on what you've done during the day. Perhaps staying up for 44 hours, flying across seven time zones, forgetting to eat any meals, partaking of all that free Champagne available on international flights, gulping down three different brands of cold medicine for the onset of a scratchy throat and then driving to the top of a 10,000-foot volcano doesn't leave you in tip-top condition. Best that you go to bed, nurse your cold and observe again some other night.

The atmosphere

The good news about Planet Earth's atmosphere is that it allows us to breathe, keeps us warm, blocks out harmful Solar ultraviolet radiation and cushions us from all but the largest meteors. The bad news is that it wavers and wiggles like a bowl of gelatin on the back of a rodeo bronco. It also supports smog, clouds, dust and, in the extreme, rain, snow and hail. Worst of all, it moves around, causing winds and it seldom appears the same twice. We can ameliorate some of these effects by moving to a mountain top or desert and thus suffer less from the effects of atmosphere and weather. But unless your 20-cm S–C is located on orbit you're going to have to deal with the atmosphere.

In general, two parameters characterize the interaction of the atmosphere with astronomy. The first is the atmosphere's transparency and the second is its stability or seeing. The transparency of the atmosphere refers to how much light is absorbed between the top of the atmosphere and you. Starlight passing straight down from the zenith loses about 0.2 magnitudes pass-

---

[68] The folklore that eating carrots will increase your vision appears to be a product of deliberate misinformation by British Intelligence during World War II. The Royal Air Force had developed night airborne intercept radar and was destroying many German planes. The existence of the radar remained a secret but a cover story was needed to explain the phenomenal change in the tide of the air war. Newsreels showed RAF pilots gulping down large bowls of boiled carrots, reputed to improve eyesight. Millions of children have since been subjected to this noxious food. I sincerely hope that this notice will end the abuse.

ing through the atmosphere to sea level.[69] If you observe with a zenith distance of 60° (30° elevation from the horizon) then you are looking through the equivalent of two air masses and thus the loss is 0.4 magnitudes. At about 70° of zenith distance (20° from the horizon) three air masses absorb 0.6 magnitudes of starlight. In general, the farther you observe from the zenith then the more starlight is absorbed.

While the 0.2 Mv per air mass rule of thumb generally works well, realize that the light from different colors is absorbed differently. Blue light is absorbed more than red. In actuality, it is scattered or reflected off small dust particles.[70] In some geographical areas, absorption may differ because of higher concentrations of natural or artificial aerosols and dust. Volcanic events can throw millions of tons of dust into the atmosphere, causing higher absorption world-wide for months or years.

While theoretically a 20-cm S–C should have a resolution of 0.6 arc seconds[71], on many nights the atmosphere and not the telescope will limit seeing.[72] The seeing or steadiness of the atmosphere is determined mostly by upper-atmospheric wind patterns. It can also be affected by localized heat sources such as a lake next to the observatory which causes updrafts or downdrafts. The time of day can also be a factor since periods of great temperature change such as just after Sunset or Sunrise can cause large thermal movements in the atmosphere. Thus, just before dawn is often a time of steady seeing. As the atmosphere undergoes turbulent motions, it forms weak lenses of denser and lighter gas which bend the incoming rays of starlight. Usually, the changes in these lenses are so rapid that the result is a blurring in the image into a fuzzy circle.

Occasionally, however, the seeing will calm down for a few seconds and exceptional views can be had. Astronomers search

---

[69] *Allen's Astrophysical Quantities,* third edition, C.W. Allen, Athlone Press, 1973, p. 127.
[70] Scattered blue light is what makes the sky so blue.
[71] An arc second is an angular measurement defined as 1/3600 of a degree. That is the apparent width of a standard MacDonald's French fry (potato chip in the UK) seen from a distance of 1 km.
[72] The limiting resolution of telescopes is discussed in *ASTRONOMY Magazine,* November, 1991, supplement p. 9. For a description of the meaning of optical resolution, see the article *Optical Quality in Telescopes* by Peter Ceravolo, Terence Dickinson and Douglas George, *Sky and Telescope,* March, 1992, p. 253. For an interesting home-made device which allows you to measure the resolution of your own telescope, see the letter by Andreas Maurer in *Sky and Telescope,* September, 1991, p. 311.

the world over to find geographic locations where this happens often. Mars Hill in Flagstaff, Arizona, is one such spot. For just a few moments now and then, superb seeing allows observations with resolutions much better than an arc second. In order to observe there, however, you'll have to ask permission of Lowell Observatory first, for Percival Lowell discovered the spot about a century ago.[73]

As with the problem of transparency, the more atmosphere you look through, the worse the conditions will become. For this reason, if you want to observe a particular object and have the luxury of picking a time for the observation then wait until it is highest in the sky for the best view. Granted, if you want to observe the Orion Nebula in the Spring then you'll have to do it just after Sunset low in the West. The alternative is to wait until Fall and observe it higher in the East.

After a while as an astronomer, you will learn to predict not only night time clouds but periods of good seeing and transparency. While most newspaper and television weather forecasters have this information, they seldom announce it. Only in Arizona, with perhaps the largest concentrations of amateur and professional astronomers, is this information regularly broadcast. It took ten years of concerted efforts by the entire astronomical community to get just one nightly astronomical forecast on the local TV news program.

The observing site

For most of us, the observing site is simply whatever back-yard space that came with the house, for few people move to a location with observing in mind – although I know several astronomers in Arizona who did just that. The first requirement is that the site have good horizons. It is frustrating to slew the telescope to some interesting object which should be well above the horizon, only to discover that it is blocked by some tree or building. My own home observing site violates this principle and often I must move the telescope and all of its accessories to the other side of the yard in the middle of an observing session

---

[73] Happily, Lowell Observatory has an open house observing session for the public at least once a week. Don't bother to bring your 20-cm S–C, though. Their 61 cm Clark refractor is a much better instrument.

just to see some particular object. This also means that I have to re-align the telescope to the pole. The alternative is cutting down all my trees and some of my neighbor's – a plan I'm sure he wouldn't appreciate. Thus, before I move the telescope, I ask myself if there are any more objects in the part of the sky that I can see that I'm going to want to view later. At least that prevents me from having to move the telescope twice in one night.

Lighting concerns are twofold; direct light and reflected light from the sky. Direct light is that from street lamps or your neighbor's back porch light which falls directly on the telescope. In addition to preventing your full night vision from developing, these lights can cause glare in the corner of your eye which distracts from viewing. Usually, just holding your hand beside your eye to block the light suffices but persistent stray light is annoying.

I once lived in a fairly dark place but when I'd take the telescope out in the back yard, my neighbor's dog would bark at the commotion. My neighbor would hear the barking, think that prowlers were about and turn on spotlights illuminating both his and my back yards. This called for a bit of diplomacy and I wound up inviting him over for a view through the telescope. He soon realized that his lights were annoying and shut them off without my directly asking him to do so.

Reflected background light from the sky is an increasing problem in most areas. There are several solutions, none of which are completely satisfying. You can ignore all of the faint objects in the sky and take up a program of Lunar and planetary observing. You can move elsewhere (but be prepared for others to move there too and re-create the original problem). You can pack up your telescope and haul it out into the countryside but this is time consuming and expensive. You can join the International Dark Sky Association (IDSA) and lobby your local government to change its lighting habits. I have participated in such exercises and we now have an ordinance that advertizing lighting for billboards must be turned off after Midnight. The passage of the ordinance and its subsequent enforcement are an interesting civics lesson but they cut into valuable observing time.

The seeing at a site is affected by local thermal sources. I know that if I look to the Northwest early in the evening, the heat rising from the concrete and asphalt of the nearby airport is going to distort the atmosphere and make star images large. Planets, fortunately, seldom set that far North so it is not a great problem. There is little that one can do to change the local countryside but one caveat is worth mentioning. I was in the back yard (South side of the house) setting up for an evening of observing. While aligning on Polaris, the seeing suddenly decreased to a level I had never seen before in my life. I'd heard of instances where observers tried to view objects at the edge of the Jet Stream, a fast, high-altitude wind current prevalent over North America. Such air movements are reputed to cause bad seeing. I became interested and started to devise some method whereby I might be able to measure quickly just how bad the seeing had become when suddenly everything calmed down, the star becoming a fine point. Confused, I went inside and got a small radio receiver tuned to the weather broadcasts to see if there were some upper-atmosphere disturbance. Listening to the radio, I sat down at the telescope again and suddenly Polaris blossomed into a huge fuzzball even larger than before. At that point I realized the cause of the problem. I was looking directly over the chimney of my house and every time the gas-fired furnace turned on, I was look-ing through a column of hot air churning upward.

Site security isn't an issue when I use the telescope at home but in the field I do not observe alone. I'm not worried about other people as much as I am about stepping on a rattlesnake – a very real danger in Arizona. While I have yet to see or hear a rat-tlesnake during observations, I still take precautions like making a lot of noise when I get out of the car to scare them off. Your observing site may have other dangers and you should consider them before venturing out. The one exception to the rule of never observing alone is when a group of astronomers goes to the field for asteroid or graze observations. There simply are not enough astronomers to afford two at each site since we want to maximize the number of sites. In such cases, we remain in contact with each other via portable two-way radios. We check in with each other every half hour or so and at the end of the night we gather at some meeting place to ensure that every vehicle has made it out of the desert.

Several of my acquaintances have bought land away from the city on which to observe. A few have built permanent observato-ries and one has retired to his remote site to live. While such a location may seem ideal, there are two warnings in order. The first is to thoroughly investigate the site for dark skies and good seeing. Major professional observatories conduct site surveys lasting years before investing millions of dollars in new facilities. The amateur should at least visit the site several times at different seasons before purchasing land. The second caution is to examine how often the site will be used. More than once I have watched an astronomer search far and wide for the ideal site, settling on some remote location with pristine conditions. Unfortunately, the site is a four- to six-hour drive away (one site was accessible only by air) and thus the location is not visited often. If astronomy is your hobby and not your profession then consider that the hours spent getting to the site constitute leisure time and thus detract from the pleasurable activity of observing.

4

Deep sky

The term 'deep sky' has been used to cover a wide variety of dissimilar objects. Generally it means objects which are neither Solar system members nor stars. Deep sky can be broken out into two broad classes; non-stellar objects which are members of our own galaxy (nebulae) and external galaxies. The study of these faint, fuzzy patches of light actually has it roots in the search for comets. While Herschel and others noted the presence of whispy images, it was Charles Messier who catalogued them. In actuality, he was searching for comets and occasionally came upon small glowing clouds which looked like comets. In order to eliminate the false comets, which do not move with respect to the stars, he made a list of non-cometary extended glowing patches. In his day it was unclear that some of the objects were internal to our galaxy while others were external galaxies in their own right. Indeed, at the time, the true structure of our galaxy was only beginning to dawn on Herschel. Thus, the Messier Catalogue has external galaxies and internal galactic objects mixed together.

Messier's list of non-comets is shown in Appendix 10. There is some question about several of the objects. At least one, M102, is an accidental repetition of M101. M40 is just a double star. The last five or so objects were actually reported by Messier's colleague, Pierre Méchain. Various lists have been published and debated in their fine detail as to whether this or that nebula should be included. In spite of this, the list remains a handy catalogue of the finest deep-sky objects which can be seen from Paris. Note that for Southern Hemisphere observers there are wondrous sights such as the giant globular cluster Omega Centauri, Eta Carinæ and, of course, the Magellanic Clouds. They just weren't observable from Charles Messier's Paris and therefore they didn't appear on his list.

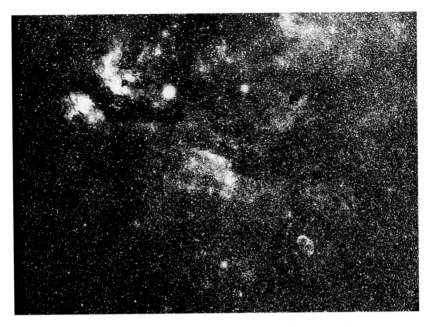

Figure 4.1. Dark nebulae near Gamma Cygni. Photo by Chris Schur.

Gas, dust and stars as nebulae

Although we may think of space as a vacuum, there exist considerable quantities of dark material between the stars. While planets, asteroids and comets make up a small portion of the mass, a much larger fraction is in the form of gas and dust clouds floating in space. These clouds may be light years in size and yet so tenuous that they contain less matter per cubic centimeter than the best laboratory vacuum obtainable on Earth. In spite of this, they have so much matter that a cloud thickness of only a light year or two is sufficient to block out all of the light from stars behind the cloud. An example is shown in Fig. 4.1, taken by Chris Schur with a 20-cm Schmidt camera. While there are occasional stars within the dark areas, these are foreground stars between the dark nebula and the observer.[74] Southern Hemisphere observers have a much closer and larger example in the Coal Sack Nebula, close to the foot of the Southern Cross.

[74] For more on dark nebulae see *Sky & Telescope Magazine*, August, 1991, p. 207.

Figure 4.2. The Horsehead Nebula. Photo by Chris Schur.

This is a naked eye cloud of gas and dust which blocks out the light from stars farther from us in the Milky Way.

One other place where gas and dust clouds produce dark patches is in the Sagittarius Star Cloud, looking towards the densest part of our Milky Way Galaxy. Some fields of view are filled with stars while some are curiously empty. A wide field-of-view eyepiece or rich-field adapter shows the shapes of dark clouds of obscuring matter. Indeed, we cannot see anything in that direction beyond about one third of the distance to the center of the galaxy due to intervening matter.

A more interesting phenomenon occurs when a nearby star illuminates the gas and dust, often making the gas fluoresce or glow. Fig. 4.2 shows the area around Zeta Orionis including not only an illuminated cloud but nearer dust clouds which obscure part of the brighter nebula. Just to the right of center an oddly shaped dust lane obscures the brighter nebula behind and forms the Horsehead Nebula.

A mere three and a half degrees South Southwest of Zeta Orionis is a similar, but much more spectacular, object. The Great Orion Nebula, also known as M42, is a vast cloud of gas and dust

Figure 4.3. Orion Nebula, 50-minute exposure at prime focus on Kodak Ektachrome 400 film. Photo courtesy of Meade Instruments Corporation.

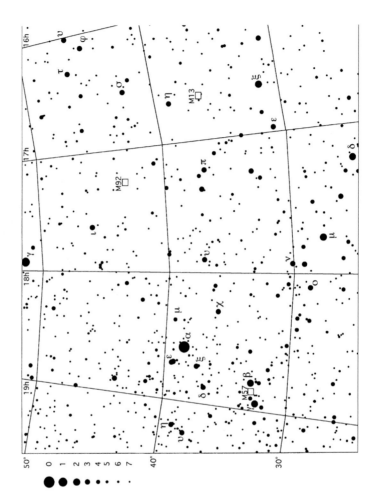

Figure 4.4 Finder chart for M13, M57 and M92.

located only 1600 light years away. While it is about 26 light years in diameter it is close enough and bright enough to be seen barely by the naked eye. Look at the middle star in the sword of Orion and with a little reflection you'll realize that it's not a star and appears slightly fuzzy around the edges. A finder chart for this finest of nebulae is shown in Fig. 2.6.

The light from several young, hot, bright stars makes a small portion of the mostly dark nebula glow. Within the cloud is a stellar nursery where stars are forming as the gas and dust collapses into small knots. Such clumps, a light year or two in size,

Figure 4.5. Photograph of the area around the Ring Nebula. The bright star on the right is Beta Lyrae and is also shown on the finder chart in Fig. 4.4.

heat up as their own self-gravity pulls them into smaller and smaller volumes. Infrared sensors which can pierce the obscuring dust have revealed many such proto-stars. Eventually, many of these concentrations will 'turn on', fusing hydrogen to helium as they mature into common stars. Some will generate a strong Solar wind which will sweep the remaining gas and dust from the area, allowing outside observers to see the previously hidden stars. In a mere twenty million years or so this area will look like an open cluster of bright stars accompanied by hundreds of fainter dwarfs, much like the Pleiades star cluster. The difference will be that the Orion cloud is huge, encompassing an area in the sky that is twice the diameter of the Moon. Better take a quick look right now before all the pretty nebulosity is gone.

While the Orion Nebula shines partly by light reflected from dust particles, emission nebulae are composed mostly of gas. Generally they are illuminated by a very hot star which shines much more brightly in the ultraviolet than it does at visible wavelengths. Such ultraviolet or 'black light' can cause gases to

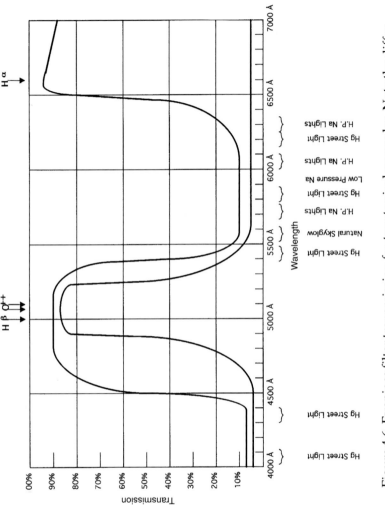

Figure 4.6. Eyepiece filter transmission for two typical examples. Note the difference between the filters at the red (longer-wavelength) end of the spectrum.

fluoresce or glow with colors characteristic of the particular gases.

A good example is the Ring Nebula, also known as M57. This interesting object was formed when, in the normal course of a large star aging, it shed or ejected a shell of material several thousand years ago. The star itself is in the center of the nebula and, at Mv 15, just beyond the abilities of most visual observers with a 20-cm aperture telescope. It can easily be photographed, though. The star shines brightly in the ultraviolet and makes the mostly hydrogen ring glow like a fluorescent light with a total brightness of about Mv 9. While the nebula is actually a nearly transparent spherical hollow shell of gas, we see it only at its outer edges because at those points we are looking through more of the shell than at any other place on the sphere. The image appears as an oval about 1 minute of arc by 1.3 minutes of arc.

Nebular filters

Many observers living in urban areas with bright skies would like to block out the reflected photons from street lamps and illuminated billboards. There are on the market many filters for this purpose but often the observing results are not as advertized. Before investing in such a filter, the observer should first consider the source of the offending light. If he lives in a community with low-pressure sodium lights (deep yellow color) then a simple filter which blocks the two narrow wavelengths of sodium emission will suffice. The loss of these two wavelengths is hardly noticeable for most regular observations and the apparent sky background will decrease dramatically.

If, on the other hand, you live where high-pressure sodium lamps (white-pink or light yellow) or mercury vapor lamps (white with a tinge of blue-green) prevail then you have greater problems. These lamps have some emission lines but they are either so crowded together that they appear like continuous spectrum or they have a true continuous spectrum. By the overall color of the lamp you can see that some of them have more photons in either the red or blue end of the spectrum. If you have incandescent lighting then its spectrum is a true continuum like the Sun's.

One solution has been to construct a filter blocking the brightest lines of the mercury streetlight, which is the commonest type of outdoor lighting in North America. It also partially blocks a continuum of light emphasized toward the blue end of the spectrum. While some of the light from nebulae will be blocked, a greater percentage of the offending light may also be blocked. If you are looking at a gaseous nebula which glows in the red light of hydrogen, then a lesser percentage of the nebula's light will be blocked than that of the sky background. The nebula will appear slightly fainter with the filter but the background will appear much darker, with a net enhancement in contrast. On the other hand, if you are looking at a dust nebula illuminated by yellow and blue stars then the filter won't help much. Similarly, if your local lighting isn't exactly that which the filter was designed for, then its usefulness is limited.

Finally, there are nebular filters which block everything except a very narrow range of wavelengths at some expected emission line color, such as the red of hydrogen. These work admirably but only on objects which emit at that specific wavelength.

Clusters of stars

When writing about the great Orion Nebula I mentioned that eventually it will look like the Pleiades, an open star cluster located near it in the sky. Fig. 2.6 shows the location of this bright grouping of stars, approximately 410 light years from us.[75] While the cluster can easily be seen with the naked eye, examination with the 20-cm S–C reveals several hundred fainter members of the cluster. When observing under dark skies with a wide-field-of-view eyepiece, a faint nebulosity can be seen around some of the brighter stars.[76] These are the last whisps of gas left over from

---

[75] Nearly every civilization which has observed it has made up legends and stories about the cluster. In European tradition it is called the "Seven Sisters". The fact that usually only six stars can easily be seen indicates that one of the stars may have faded within the last 1000 years, not an unusual occurrence. In Western Pacific cultures it is called "The Necklace" and is associated with a popular folk tale in which a hunter returns home to find his family dead. In his grief he flings a necklace he has made for his mate into the sky and the necklace becomes the loop of stars we see today. The word for necklace in Japanese is Subaru and a stylized emblem of the cluster appears on the trunk (boot) logo on an automobile.

[76] Near the star Merope is the brightest of these patches, also known as the nebula NGC 1435.

Figure 4.7. The Pleiades Cluster with Mars. Photo by Chris Schur. Note that the photo was taken with a 20-cm Schmidt camera.

the formation of the cluster. They are illuminated by the hot, bright main stars of the cluster. Fig. 4.7 is a photo from Chris Schur's 20-cm Schmidt camera of the Pleiades. The bright object at the bottom of the photo is the planet Mars.

Between the Pleiades and Orion is another more open cluster called the Hyades in the constellation of Taurus. While you may or may not agree that the stars look associated, careful studies of the proper motion and radial velocity of the cluster members reveals that they are all moving in the same general direction at the same general speed. This implies that they all formed at about the same time and place, probably in a large gas and dust cloud like the Orion Nebula. While this open cluster spans almost 20° of sky, there is at least one obvious star which does not belong to the cluster. The bright red Mv -0.7 star Aldebaran (α Tauri) is only 68 light years from us while the Hyades Cluster members average about 130 light years distance.

Southern Hemisphere observers can look to an excellent open cluster called the Jewel Box, near the Southern Cross at about

12h51′ and -60°. Naked-eye observers may mistake the star Kappa Crucis for a single point but when the telescope is turned on it, about 100 stars of various colors become visible. While much smaller than the Pleiades Cluster, the Jewel Box is undoubtedly one of the finer sights in Southern skies. A similar distant and thus smaller cluster in the North is M44, the Beehive Cluster[77].

There exists at least one open cluster closer to us but it is very difficult to distinguish, although individual members are easy to see. The cluster, often called the Ursa Major Group, includes the stars ε, β, ζ, γ and δ from that constellation. Other prominent stars with similar motions with respect to the Galaxy include α Ophiuchi, δ Leonis, β Aurigæ and α Canis Majoris (Sirius).[78] Thus, the cluster almost surrounds us but our own Sun is not a member. We are just passing through the cluster for the next few millennia. Our inability to discern the outlines of the cluster is analogous to being unable to determine the shape of the forest because we are inside it. From the outside, this cluster might look quite pretty, even if the view is marred by that one small yellow type G star with the nine planets passing through.

Globular clusters

While open clusters are characterized by hundreds of stars, globular clusters have thousands or even hundreds of thousands of stars. This huge concentration of mass implies that the globular cluster stars are gravitationally bound to each other while most open cluster stars will eventually drift away from each other. As globular cluster stars orbit a common center of mass, the overall shape of the collection tends to become spherical.

Globular clusters are a particularly nasty problem for comet hunters since they look almost exactly like a distant comet with a well-developed coma but no tail, a common comet appearance when the object is far from the Sun. Thus, Charles Messier marked many of them on his list of non-comets. The brightest Northern globular is M13 in the constellation Hercules, as shown in the finder chart in Fig. 4.4.

---

[77] The Beehive Cluster has alternatively been called the Praesepe.
[78] *The Sky: a User's Guide*, David H. Levy, Cambridge University Press, 1991, p. 197.

Figure 4.8. Globular cluster M13, 25-minute exposure at prime focus with Kodak #2415 film. Photo courtesy of Meade Instruments Corporation.

Southern observers have even better examples in Omega Centauri and 47 Tucanæ which look like fuzzy Mv 5 stars.[79] These are the best globular clusters visible from Earth and well worth seeing even if you have to get up at some odd hour to find them.

In a 20-cm S–C telescope globular clusters appear as a fuzzy circle, sometimes flattened a bit into an oval. A little inspection may reveal individual stars on the outskirts of the cluster if the seeing is steady. On excellent nights brighter stars nearer the inner regions can be distinguished. Such views challenge the resolution limits of your telescope and in some of the smaller clusters, individual stars can be discerned all the way to the center.

There are about 120 to 150 such clusters within our galaxy. They tend to form a rough sphere centered on the plane of the

[79] Omega Centauri, located at about 13h24' and -47°, will rise barely above the horizon for observers in the Southern US. While it usually is distorted and wavering as seen through the thick atmosphere, it is impressive in brightness and size, being nearly as large as the Moon in angular extent.

Figure 4.9. Omega Centauri, 10-minute exposure at f/6.2 on
Fujichrome RD 100, gas hypered. Photo by Bernd Koch and Jörg
Stahlhut, courtesy of Celestron International.

Milky Way. Thus, as we view them from our location within the
plane of the galaxy, most globulars appear far from the Milky
Way in darker, less-crowded regions.

Supernova remnants

Only rare, massive stars can destroy themselves by exploding
but when they do the result is a spectacular event. On July 4,
1054, Chinese astronomers recorded a 'guest star' in the constellation Taurus. It became as bright as Jupiter and remained visible
to the naked eye for two years. The central portion of the star collapsed into a dense neutron star while the rest was flung out-
ward in a chaotic explosion. Nearly a thousand years later we
can see this material as the Crab Nebula, also known as M1. A
finder chart is in Fig. 2.6.

The nebula has expanded to about six light years in diameter
which makes it about 180 arc seconds in diameter at its distance

of 3500 light years. It is still growing in size at a rate of 700 km/s or about 0.2 arc seconds per year so small changes will become apparent over a lifetime of observing.[80] The intricate system of filaments and knots of hydrogen, helium and other elements glows brightly. Power to illuminate the nebula comes from the rapidly spinning neutron star at the center. Its magnetic field creates a natural synchrotron, sweeping up electrons which emit photons when accelerated. This also ionizes the gas which glows at its characteristic emission lines. The magnetic field of the spinning neutron star also emits radio, X-ray and optical pulses at a rate of 30 times per second.

Supernova remnants do not last long with respect to the lifetime of the galaxy. They tend to fade rapidly and in a matter of just a few hundred thousand years they dissipate. The Veil Nebula in Cygnus and, for Southern observers, the Gum Nebula are examples of older supernova remnants. Only two other supernovae have been observed in our own galaxy since the 1054 event. In 1572 and 1604 supernovae were observed by Tycho Brahe and Johannes Kepler, respectively. Estimates of supernova production range from one to three per century per galaxy. Obviously more have occurred recently within our galaxy but they were probably obscured by dust within the Milky Way.

We often see supernovae in external galaxies. External supernovae are usually so distant and faint that their light is difficult to study but many are well within the grasp of a 20-cm S–C telescope. Furthermore, they are sometimes not noticed immediately and thus the more interesting early phases of the event are missed.[81] In 1987 a supernova in the nearby Large Magellanic Cloud was observed intently by Southern Hemisphere astronomers.

---

[80] The debris are moving fast enough to escape into space and thus fuel the next generation of stars with a mix of gas and dust enriched in heavier elements. See *Amazing Universe*, Herbert Friedman, 1975, National Geographic Society, pp. 99 & 105.

[81] A dedicated amateur with a 20-cm S–C telescope can patrol several galaxies per night to look for "new" stars. If you observe 30 galaxies per night you would expect to see about 1 supernova per year. Professional astronomers rely on amateur teams to find these events quickly and report them so that more-powerful spectrographs can be turned on the new event.

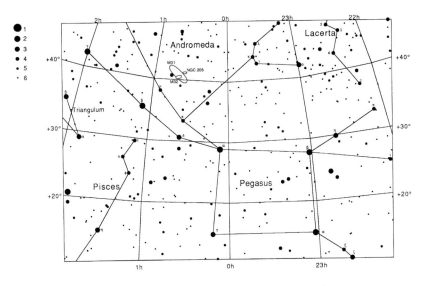

Figure 4.10. Finder chart for Andromeda Galaxy.

Find a galaxy

Before you ever look at a galaxy you should be warned about one of the great disappointments in amateur astronomy. In a 20-cm S–C telescope, galaxies aren't anywhere near as spectacular as those photos on the covers of popular astronomy books and magazines. Those photos were taken with very large telescopes by professionals or very competent amateurs using state of the art equipment and every trick known in the darkroom or computer image processing lab. If you lower your expectations a bit then the real image of a galaxy through your own telescope will be interesting not so much in its spectacular detail as it is amazing that you were able to find a galaxy by yourself.

Galaxies are vast clumps of a billion or more stars gravitationally bound to each other. Many astronomers consider galaxies to be the basic building blocks of the Universe. They come in all shapes and sizes, the most familiar being a pinwheel shape like our own Milky Way Galaxy. Even more of them are ellipsoidal in shape with fuzzy edges. Others are irregular patches of stars.

One of the nearest large galaxies is a pinwheel type slightly larger than our home galaxy, located in the direction of the constellation Andromeda.[82] On a good night in dark skies you can just barely see it with the naked eye. This is probably about the farthest object, at 2.2 million light years, you'll ever see without a telescope. While the light reaching us is 2.2 million years old and thus we see the galaxy as it was in our own pre-history, its shape won't have changed much in the intervening years since this duration amounts to only about 1/100 of its rotation period.

The first impression of the Andromeda Galaxy, also called M31, in a telescope is of a slightly elongated fuzzy sphere.[83] A wide-field-of-view eyepiece shows details best. Various estimates have been given for its size since the edge of the galaxy is not a clear line. Farther from the center the stars thin out and larger-aperture telescopes show a much larger galaxy. It is at least three times the apparent size of the Moon and photographs have been taken of it with extremely fast (low f ratio) satellite tracking cameras showing it to be 4° to 6° across.

The bright fuzzy patch at the center is only the core of the galaxy. With a little careful observation and scanning the telescope around gently one can see the disk surrounding the core. Then some of the more prominent dust lanes near the core will become evident. This is an object which requires a little time to see all of the details. It is larger than the field of view of most eyepieces. Months or years later you will still be discovering new features in this fascinating object. It is, after all, a complete galaxy including spiral arms, globular clusters, glowing regions of hydrogen gas and the occasional nova or supernova.

With a little sweeping around the nucleus, two smaller dwarf galaxies come into view. They have been designated M32 and NGC 205. These and at least six other smaller dwarf galaxies are gravitationally bound to Andromeda. Most large galaxies have a swarm of smaller dwarf galaxies hanging around them. M32 is located about 24 minutes of arc South and a little East of the core of the main galaxy. NGC 205 is about 35 minutes of arc Northwest of the core. Another scenic point of interest is NGC 206, a cluster of young, hot stars and dust located in a spiral arm of the main Andromeda Galaxy about 50 minutes of arc Southwest of the core.

[82] An interesting article on how to observe Andromeda and its companion galaxies appears in *ASTRONOMY Magazine*, November, 1991, p. 76.

[83] The Andromeda Galaxy is often referred to as the Andromeda Nebula, a misnomer left over from the days when all extended fuzzy patches were called nebulae.

Figure 4.11. Whirlpool Galaxy, a typical spiral type; 35-minute exposure at prime focus with Kodak #2415 film. Photo courtesy of Meade Instruments Corporation.

Figure 4.12. M83, a barred spiral galaxy; 55-minute exposure at prime focus on Fuji 1600 film. Photo by Bruno Mattern, courtesy of Celestron International.

More galaxies

Southern Hemisphere observers don't need finder charts to locate the nearest galaxies. They don't even need a telescope. The Large Magellanic Cloud and the Small Magellanic Cloud are both dwarf irregular galaxies which are probably gravitationally bound to our own Milky Way Galaxy.[84] A score or so smaller dwarf galaxies hang out near our galaxy, just as Andromeda has a set of smaller companion galaxies. Most of these others, however, have very low surface brightnesses and are difficult to see. Still, it is fortunate that both of the Magellanic Clouds are bright and easy to study.

As with any science, astronomers have classified galaxies into types. The three main types are spirals , ellipticals and irregulars. That third category might also be labeled 'miscellaneous', for it contains objects which simply don't fit any convenient classifica-

---

[84] Some galactic astronomers classify the larger cloud as a barred spiral.

Figure 4.13. NGC 253, a spiral galaxy seen nearly edge-on; 55-minute exposure at prime focus on Fuji 1600 film. Photo by Bruno Mattern, courtesy of Celestron International.

tion scheme. Spiral galaxies are lens shaped. In general, they have a bright central core from which two or more spiral arms emerge which wind around the center. These galaxies can be further subdivided into categories describing whether the spiral arms are loosely or tightly wound. Some spirals also appear to have a bar across the center which connects the arms, as shown in Fig. 4.12.

In some galaxies the spiral arms and dust lanes are well defined while in others the disk of the galaxy appears to be a featureless sheet made up of evenly spaced stars. In most of them, though, there is some differentiation which indicates that the whole mass of the galaxy is rotating like a pinwheel. While spiral galaxies have much greater diameters than their thickness, how they appear to us depends on the angle between their axis of rotation and our line of sight. The galaxy in Fig. 4.11 is seen nearly face-on. Many spiral galaxies are seen edge-on and thus they appear to be just a fuzzy line in the sky. A typical example is shown in Fig. 4.13. Some spirals seen edge-on also exhibit a relatively thicker central core with a much thinner disk.

Figure 4.14. NGC 5428, an elliptical galaxy with a dust lane; 55-minute exposure at prime focus on Fuji 1600 film. Photo by Bruno Mattern, courtesy of Celestron International.

The second major galaxy type is the elliptical. These are egg-shaped collections of stars that rotate like spiral galaxies but do not show the complex structure of spirals. Some of them, however, exhibit a dark dust lane across the middle, such as NGC 5428, as shown in Fig. 4.14. Some elliptical galaxies are very elongated while others are nearly spherical.

There exists a 'cosmic zoo' of irregular galaxies, colliding galaxies, peculiar galaxies and some with exploding cores. Indeed, it has been theorized that some galaxies may contain a black hole at the center, massing many millions of Suns. There even exist clusters of galaxies such as the concentration in the constellation of Virgo including M84 and M86. Another example of a tightly knit group is Stephan's Quartet, a group of four associated galaxies in the constellation Pegasus.[85]

[85] The group was once referred to as Stephan's Quintet but recent measurements of radial velocity have suggested that one of the members of the group is not gravitationally associated with the others. Still, they make a pretty sight. See *Travellers in Space and Time*, Patrick Moore, Doubleday, 1984, p. 167, and *Galaxies*, Harlow Shapley revised by Paul W. Hodge, Harvard University Press, 1972, p. 157.

The Messier Marathon

With such a plethora of extended objects it might be difficult on any given night to make up your mind as to what to look at. Should you go out and enjoy familiar nebulae or work at the craft of observing and search for ever fainter galaxies? The choice is entirely yours but you will soon find that a little planning before observing helps the session go better. Tactics come into play when observing objects during the early evening in the Western sky. Those objects which will set first should be observed first. Those objects which are still rising in the East can wait until later in the night – unless the Moon will rise later, in which case its light may wash out subtle details.

The question arises as to whether it is possible to observe all of the Messier objects in a single night. One would have to chose a date when the Sun were not near any of the Messier objects. A quick scan of a map of Messier object locations reveals that they are well spread out over the Northern skies. There is, however, a gap from about 22 hours to 1 hour of right ascension with only a couple of objects at high Northern declinations which makes them visible almost all night. During March and early April the Sun traverses this gap.

Many amateur astronomers have chosen the best Moonless night during this period, armed themselves with finder charts and strategy sheets and attempted a Messier Marathon. Usually the event is celebrated as an astronomy club star party[86] with each participant helping others find the more difficult faint fuzzy patches. The evening begins with a desperate search for M74 and M77 as they sink into a twilight haze. Then more accessible objects can be viewed. Planning and prior practice in finding objects pay off as each item is logged.[87] Towards the middle of the night observers slow down as they pick their way through the dense Virgo cluster of galaxies with more than a dozen Messier objects and another dozen fainter galaxies visible in a 20-cm S–C within a ten degree square.

---

[86] I dislike the term "star party" and prefer "observing session". The reason is that about once a year at one of our club's star parties, somebody shows up spraying white light all over the site while looking for the beer and the band. It's not that kind of party.

[87] A suggested search strategy is given in *The Sky: a User's Guide*, David H. Levy, Cambridge University Press, 1991, p. 221.

If the coffee holds out and the observer can stay warm and avoid fatigue then well before the first hint of dawn, all but the last few Messier objects will have been logged. Now comes the wait for M30 to rise above the horizon. The first brightening of the horizon precedes even the finder stars you've plotted just West of M30. You worry about clouds in the East and hope that it won't rise behind the one distant tree on your horizon. Tension mounts. Will you bag that last little globular cluster? Go find out.

The observer's log

Several times I have mentioned keeping a log of observations. While the log is primarily for the observer it can also aid other astronomers. Even innocent notes by an observer ignorant of what he sees can become valuable. The notation, '10:20 p.m. There's a pretty foreground star in front of M101 that I'd never noticed before', might actually be the first observation of a supernova in this nearby galaxy. Although the neophyte doesn't know it yet, he has valuable information on when the event occurred. Often it might be days before another astronomer notices and realizes the significance of that 'foreground star'. By that time it might be unclear as to whether the supernova is three days old or ten days old. Professionals often ask the amateur astronomy community for information on pre-discovery observations of such things. Realize that there are only a couple of hundred large professional telescopes in the world. Almost all of them have a field of view much less than one square degree. There are 41 253 square degrees in the sky and every one of them has several faint galaxies, quasars or stars capable of producing an outburst or change of some kind. Amateur observers thus have the opportunity to make basic discoveries.[88]

The observing log is mostly for the observer himself. About once every two years I record a total Solar eclipse. A decade ago I

---

[88] It has been observed by several professionals that astronomy, of all the sciences, is the most democratic. If you are going to study anthropology or mineralogy then you must join an institution which possesses a good collection of specimens. Only a handful of scientists can gain access to world-class specimens such as the oldest human remains or the Hope Diamond. In astronomy, however, all of the objects in the sky are visible for free to anybody who wishes to look. Granted, you may have to spend a little money to obtain a telescope but you already own a 20-cm S–C. Thus, the estimated 100,000 owners of 20-cm telescopes have a better instrument than 99.9983% of the rest of the people in the world.

spent two whole nights finding the correct combination of relay lenses and mechanical supports to record an object the size of the Moon or Sun with good resolution to the edge of the frame and enough stiffness in the camera mount. Then I spent a hot, sweaty afternoon in the August desert heat determining the correct filters and exposures to use on the Sun. I'll never have to do that work again and each time I record an eclipse I'm glad I took good notes the first time.[89]

Most of your log will consist of lists of objects observed and your impressions. Don't leave out notes on equipment and accessories, though. Record eyepieces, cameras and other paraphernalia used. My observing log resides in a text file on a computer so it's easy for me to find the last time I observed M33, without having to reread the whole log, by doing an automated key word search. Thus, when I reviewed notes of my first observations of Orion using my then new 20-cm S–C I was puzzled as to why I hadn't tried my favorite 25-mm Erfle eyepiece. A quick search revealed that I hadn't owned that optical gem until two years later. Consequently, I paid less attention to the earlier Orion observation notes than the later ones using better equipment.

Recording the local weather convinced me after several years that trying to observe after about 3:00 a.m. local time for two months of the year was useless unless you were willing to spend about half your observing time fighting dew. I hadn't noticed any correlation among observing sessions because only ten incidents were spread out over the three or four years. Before I spent time and money designing an anti-dew heater I searched the logs to see how bad a problem it was. That's when I noticed a correlation in dates. A little hunting also revealed that there were no early morning observing sessions during those months when I was not bothered by dew. Although I never got around to building an anti-dew heater for my corrector plate, I now approach early morning observations in November and December only when armed to the teeth with a front dew shield, foam covers for the telescope and my daughter's hair dryer.

---

[89] The observer's log is also a nostalgia instrument. While researching this book I went back into old logs to remind myself how to perform some of the more basic observations. Luckily, I started this review on a cloudy evening, for I discovered a flood of old memories, notes about observing companions who became lifelong friends and incidents I'd nearly forgotten about. It's quite a trip down memory lane.

Your use of an observing log will depend on your own style and needs. I've given some suggestions on what I put in mine. When in doubt about what to include, I err on the side of caution and write it down. You see, I have this fantasy that a hundred years after I'm gone some researcher will pick up my dusty log, see some anomalous sketch or observation and say, 'Hey, this guy didn't know it but he observed that lost asteroid we've just recovered. Let's name it after him!'

Sketching images

For most of the history of astronomy, the human eye has been the only available detector and a hand-drawn sketch the only recording medium available. Only in the last century have photography and electronic devices come on-line. The art of rendering an accurate representation of an astronomical scene is not dead, however. It lives on in the amateur community as a recreational pastime and for some planetary study programs as a useful scientific tool.[90]

Unfortunately, in my house, all of the artistic talent seems lodged in my children who draw well. I, on the other hand, can't render a straight line without a ruler on either side of the pencil. I can, however, pass on a few tips from those who attempted to coach me prior to writing this section of the book.

There are a variety of artistic styles and media used by astronomers. Three of my coaches suggested that the beginner look at several sketching styles and experiment. Whatever feels good and produces satisfying results is the right method for you. The fourth coach insisted that there was only one way to sketch astronomical objects and that was by his method. I shall not be passing along any more of his advice.

The first object to practice on is logically the Moon. It is large in the eyepiece, provides a variety of landscapes and, for some of the maria, shows subtle changes in shading which the artist must render faithfully. It provides both large-scale and small-scale

---

[90] Until CCD cameras and sophisticated image processing tools reach the hands of thousands of observers, visual sketching will remain a prime tool of amateur planetary observers. Regular planetary patrols via amateur sketches are often used by professionals to fill in gaps in coverage caused by weather and instrument problems on larger telescopes.

NGC 2392, 160 X, C-8
1/19/85, Wally Brown

Figure 4.15. Clownface or Eskimo Nebula, NGC 2392 sketch by Wally Brown using a 160X eyepiece of this excellent planetary nebula.

detail which will keep the hand busy for many hours. Since the lighting on the Moon changes from night to night, one exercise might be to sketch the same area on consecutive nights, noting the changes in shadow lengths and shading.

One of my coaches first sketches in the outlines of his subject using a hard lead pencil to define boundaries between features. The fine gradations of light and darkness are usually best rendered in charcoal or some other soft lead. Once the pigment is laid down coarsely, a finger can then be used to smear the coloration into a small area of uniform or graduated shading. He makes the blending of pigment look easy but I have tried it and come to the conclusion that this is an art, not a mere mechanical skill. Of course, using soft leads means that the rest of the drawing is subject to smearing when it is handled so a clear spray-on coating of fixative must be applied after the drawing is complete.

After mastering the Moon (or at least adequately understanding the principles involved) the next step is to sketch the planets.

M-79, 100 X, C-8
1/19/85, Wally Brown

Figure 4.16. Globular cluster M79 sketch by Wally Brown using 160X eyepiece.

They are smaller and require more concentration on fine detail. In addition, the subtle shading differences on Mars, Jupiter and especially Saturn will challenge any sketcher. My coaches warn that one common error is for the artist to place too much contrast on faint features. It is, of course, natural to overemphasize small gradations in shading since these are the data that beg to be captured and higher contrast makes the drawing easier to interpret visually. The higher contrast, however, is not how the planet actually appeared.

This brings us to the question of art versus science. If the drawing is esthetic in nature, destined to be hung on a wall and admired, then some artistic license is allowed in exaggerating shading and features. It is meant to please the eye. If, on the other hand, it is meant to be submitted to the Association of Lunar and Planetary Observers (ALPO) as part of a data set recording the apparition of the planet then contrast accuracy is more impor-

M-82, 100 X, C-8
1/19/85, Wally Brown

Figure 4.17. Irregular galaxy M82 sketch by Wally Brown using 160X eyepiece.

tant. If the interested observer didn't notice some obscure feature on the planet until he'd looked at it for a full minute then somebody else viewing the sketch shouldn't notice that same detail until he has spent a comparable amount of time.

After the planets comes deep-sky sketching. Comets, nebulae and galaxies all show extremely low-contrast detail at nearly impossibly low light levels. Far from discouraging some artists, however, they take these objects on as a challenge. I have watched one observer spend ten minutes at the eyepiece before adding one or two pencil strokes to a drawing. In the end, the sketch has detail which I find impossible to see visually – and yet, if I compare the rendering to some photograph taken with a giant telescope, I do see the detail as drawn.

5

A Couple of Interesting Problems

Asteroids

Asteroids are more properly called Minor Planets, for they orbit the Sun just as the Earth, Jupiter or the other planets do.[91] They are, however, much smaller with the largest being only about 834 km in diameter. They are too tiny to resolve surface details with Earth-based telescopes and thus they appear as points, just like the stars. This does not, however, make them uninteresting, for they move amongst the stars with complex motions just as the major planets do. Occasionally they will even pass in front of a more-distant star.

The first asteroid was discovered by G. Piazzi from the island of Sicily on the first evening of the nineteenth century. He noticed it because over a period of several minutes it moved with respect to the stars. The object was named Ceres and found to be in an orbit about 2.8 times as large as the Earth's orbit. This placed it roughly between the orbits of Mars and Jupiter. The following year a second asteroid, Pallas, was discovered and by 1891 some 322 asteroids had been catalogued. Most orbited between Mars and Jupiter. Up to this point all discoveries had been via visual methods but then Max Wolf at Heidelberg applied the new technology of photography to the hunt and since then hundreds have been found each year. Astronomers keep track of the brightest 5000 or so objects regularly.

While the majority of asteroids remain between the orbits of Mars and Jupiter, there are several significant groups which stray elsewhere in the Solar system. The first of these are the Trojan asteroids. Mathematician J.L. Lagrange had predicted that an object roughly in the same orbit as Jupiter but travelling 60° ahead of the massive planet in its orbit or 60° behind it would

---

[91] One of the best references to asteroids I have seen is *Introduction to Asteroids*, Clifford J. Cunningham, Willmann-Bell, Inc., 1988, ISBN 0-943396-16–6.

remain in a stable position, forming an equilateral triangle with Jupiter and the Sun. At the time of Lagrange's prediction there were no such examples known but in 1906 the asteroid Achilles was discovered in the predicted place. A score or more of these Jovian hangers-on are now known.

A second class of asteroids outside the main asteroid belt is the Apollo group, named after the first to be discovered. These asteroids populate the inner Solar system with eccentric orbits which often lie closer to the Sun than the Earth. Some approach the Sun more closely than Venus and then recede out to the orbit of Mars. While their distance from the Sun will occasionally equal that of Earth, they are in little danger of striking our planet, as most have orbits inclined to the Ecliptic and thus they pass above or below the Earth's orbit when at our distance from the Sun. Several of these objects have comet-like orbits but they do not exhibit the gas and dust tails characteristic of comets. One theory holds that they may be burned-out comets which have shed all of their volatiles.

Asteroids such as Chiron have been detected as far as 19 AU[92] from the Sun, well beyond the orbit of Saturn.[93] Doubtless, there are more distant ones but these tiny chunks of rock at such distances are inherently faint and difficult to find. Some astronomers have theorized that the tinier moons of the outer planets may be asteroids captured by the massive gas giant planets, each of which have a dozen or more moons.

The asteroids are thought to be largely chunks of rock with perhaps a little ice coating some of them. Spectroscopy has revealed several distinct classes of surface reflection colors, giving rise to theories that the asteroids were once part of a planet which somehow broke up and some of the chunks are former surface pieces while some are former core pieces. On the other hand, some theorists also speculate that during the formation of the Solar system, the gravitational perturbation of nearby planet Jupiter disrupted the accretion of material in the asteroid belt, preventing any sizable planet from forming there.

While we know the numbers and distributions of the larger asteroids, as we look at smaller diameters, the numbers of objects

---

[92] The Astronomical Unit is defined as 149,597,900 km, the Earth–Sun distance.
[93] Object 1991DA may be slightly more distant but additional measurements are required.

become larger. We cannot see the tiniest asteroids simply because they are too faint. Recent searches for near-Earth asteroids, however, have netted objects in the two to ten meter size range which have approached the Earth closer than our own Moon.[94] This is an area where the separate disciplines of the study of asteroids and the study of meteors merge.

Finding asteroids is not particularly difficult since the larger ones frequently appear as bright as Mv 7 at opposition. Charts showing their locations as a function of time can be found in any of the popular astronomy magazines. Positions of the brighter ones can be found on many computer bulletin boards and in the Astronomical Almanac. It is interesting to follow an asteroid through a crowded star field, watching over a period of hours as it threads its way between stars.

You might compare the asteroid brightness to several nearby stars and observe that during the night the asteroid may change in brightness. This is because these objects usually rotate with a period of several hours. One side may be colored darker than the other. An alternative explanation is that many of them are irregularly shaped and thus, for an elongated object, we may observe the object first side-on and then end-on. The Association of Lunar and Planetary Observers (ALPO) coordinates visual observers of asteroids who make brightness estimates of asteroids as a function of time and thus determine their periods. Several advanced amateurs and many professionals measure asteroid brightness as a function of time with multicolor photoelectric photometers. Since there are literally hundreds of asteroids within the magnitude limits of a 20-cm telescope, amateurs can contribute useful observations on these objects.

One of the more interesting observations of asteroids can be made when one of them passes in front of a star. If the star is sufficiently bright then it seems to disappear for a few seconds as it is eclipsed by the fainter asteroid. If the time of the disappearance and reappearance are noted, then using the known distance and orbital velocity of the asteroid, its size can be determined. With several observers stationed a few kilometers apart, several slices or chords of the asteroid can be taken and thus its shape

---

[94] Near Earth Asteroid Densities And Their Detection Before Impact, Peter L. Manly, *Journal of Practical Applications in Space*, Volume 2, Issue 1, Fall, 1990, p. 33.

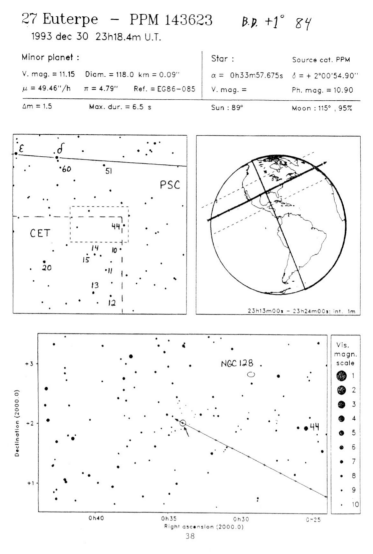

Figure 5.1. Asteroid occultation prediction data. Illustration courtesy of the International Occultation Timing Association (IOTA).

determined. The International Occultation Timing Association (IOTA) distributes predictions of these events which may happen half a dozen times a year for an observer in any given location.

Fig. 5.1 shows the data provided by IOTA for an asteroid occul-

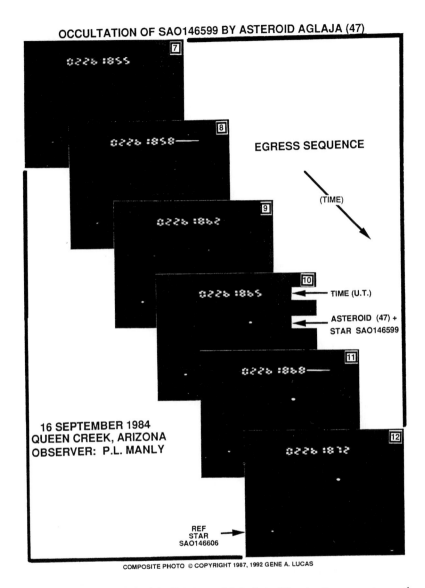

Figure 5.2. Occultation by the asteroid Aglaja. Illustration courtesy of Gene Lucas.

tation. In addition to the large scale and small scale finder charts, a view of the Earth as seen from the asteroid is provided. This is handy in judging how far the asteroid is above your local horizon. The diagonal line with the tick marks is the predicted path of the asteroid shadow and the dashed lines indicate the uncertainty in that path. All observers within the dashed lines should give the asteroid a try. The predicted time of the occultation is given in the same box as the Earth view along with the interval corresponding to the tick marks. Since the shadow takes several minutes to cross the Earth, observers in widely dispersed locations must start and stop their observations at different times.

Fig. 5.2 shows a video record of the occultation of Mv 8.7 star SAO 146599 by asteroid Aglaja on September 16, 1984. As a result of this and several other observations, the diameter of Aglaja has been determined to be 136.4 ± 1.2 km. See Fig. 11.1 for a cross-section of this minor planet.

There are groups of asteroid shadow chasers who roam the country setting up their telescopes for just a few minutes of observation during an occultation. While the predicted time of the event is usually accurate to within one minute, there is the possibility of making a secondary discovery if one observes just a little longer. It has been postulated that the larger asteroids might have sufficient mass to gravitationally capture a smaller companion object. One might call these moons of asteroids. Nobody has ever observed one of these directly but during asteroid occultations some observers far from the shadow path have seen small disappearances of the star behind something. One obvious explanation is that they simply blinked or a bird flew in front of the telescope, for the occultations generally lasted less than a second. There is, however, at least one photoelectric record and one video recording (taken with a 20-cm S–C telescope) showing secondary occultations.

In order to catch any potential asteroid moons I generally observe for ten minutes before and ten minutes after the predicted occultation time. This means that I must keep my eye glued to the eyepiece for twenty minutes without looking away. It took only one occultation, hunched over with an aching back, to convince me that I needed a comfortable observing chair and an eyepiece with good eye relief.

Although the position of the asteroid can be measured to a high degree of precision, there is always some uncertainty in the position of the shadow path on the Earth. This means that the predicted occultation path may be in error by as much as two or three times its own width. For this reason, when groups of

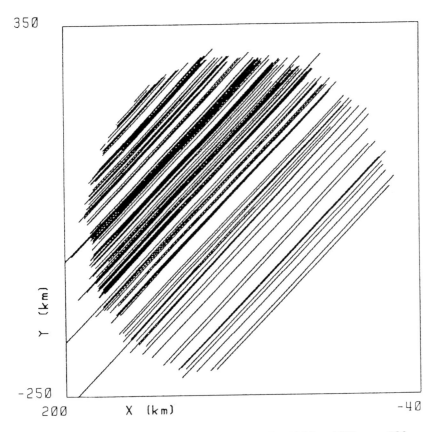

350

Y (km)

-250

200   X (km)   -40

Figure 5.3. The shape of asteroid 2 Pallas. On 29 May 1983 over 130 observers watched as the asteroid occulted star 1 Vulpeculae, allowing a measurement of the mean diameter of 563±6 km.[95] Illustration courtesy of the International Occultation Timing Associastion (IOTA).

astronomers mount an expedition to watch an asteroid occultation they spread out on either side of the predicted path. At times hundreds of amateurs have joined in such observations and measured the size and shape of an asteroid.

When observing an asteroid occultation set up early and identify the target star from the finder charts provided by the expedition organizer. Have all of your recording equipment handy and get comfortable. While you usually won't be able to see the aster-

---

[95] The Size and Shape of (2) Pallas from the 1988 Occultation of 1 Vulpeculae, David W. Dunham *et al.*, *The Astronomical Journal*, Volume 99, Number 5, May, 1990, p. 1637.

oid itself, for some events it is bright enough. Then you can watch it merge with the star until it is too close to discern as a separate object. Usually this occurs a quarter hour or more before the event. Having tested the recorder, proceed with the observation. When the star disappears, however, it will seem incredible. Many observers simply forget to mention anything on the tape at this point and thus lose the data. Get ready for the star to reappear, for most asteroid occultations last only a few seconds. Once the observation is complete be sure to record your location and all of the other data requested by the expedition. Don't wait until you get home to finish this vital part of the observation. Even if you get a 'miss' this is good data. They tell where the asteroid isn't. If the observer next to you got a hit then you will have set an upper limit on the size of the asteroid by not seeing an occultation.

Variable stars

While the stars may seem to be constant in brightness, many slowly become brighter and dimmer over a period of days, weeks, months or years. There are several possible reasons for this. Some stars are evolving, growing older and either turning on new energy sources or shedding their outer layers into gas and dust nebulae which may block the starlight later. Other stars oscillate in brightness and size during one episode in a complex aging process. Indeed, almost every star yet examined, including our own Sun, has shown minute variations in brightness when examined with instruments of sufficient accuracy.

One of the more interesting variable star types is the eclipsing binary. Remember that there are more double stars than singles like our Sun in the galaxy. Two stars may orbit each other in such an orientation that we lie in the plane of their orbit. Thus, once each orbit, one of the stars will pass in front of the other, blocking its light for a time. Depending on the relative sizes of the stars and how accurately we are lined up with the plane of their orbit, a second occultation may be seen during each revolution as the first star passes behind the second.

Usually the stars are so close to each other and so distant from us that we cannot resolve two separate sources of light and together they produce only one starlike image whose brightness is the sum of the two stars. A naked-eye example of this

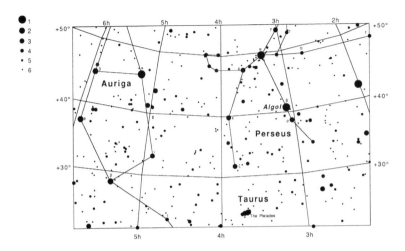

Figure 5.4. Finder chart for Algol, an eclipsing binary star.

is the star Algol in the constellation Perseus.[96] It is also called Beta Persei. For about 85% of the time it shines at Mv 2.2 but every 2.87 days it dims down to Mv 3.4 for about five hours.[97] The times for Algol eclipses are usually given in the popular monthly astronomy magazines.[98] A finder chart is given in Fig. 5.4 showing some visual magnitudes of nearby stars which will allow observers to estimate Algol's brightness by visual comparison.[99]

The estimation of star brightnesses by visual comparison with known nearby stars is almost a separate branch of astronomy, practiced by hundreds of amateurs and professionals world wide. The data are graphed into a light curve showing star

[96] Geminano Montanari, a professor of mathematics at the University of Bologna, first pointed out in 1669 that Algol varied in brightness. In 1783 John Goodricke measured the light curve and concluded that Algol is actually two stars orbiting each other periodically. See *The Picture History of Astronomy*, Patrick Moore, Grosset & Dunlap, 1961, pp. 188–9

[97] There is a smaller secondary occultation once each orbit but since the system is composed of one bright, small star and one faint, large star, when the faint star is behind the bright one the overall dimming is negligible and usually can't be detected by eye.

[98] See *ASTRONOMY Magazine*, November, 1991, p. 65.

[99] Algol is really a triple star. In addition to the two stars which eclipse each other, there is indication from spectroscopy that a third, faint companion circles the two eclipsing stars. See *Astronomy*, seventh edition, 1964, by Robert H. Baker, D. Van Nostrand Company, Inc., p. 413.

brightness as a function of time. Stellar astronomers use these curves to develop theories as to why the stars vary in brightness. While the theorists are largely professional astronomers, they use data generated by amateurs.[100] In Chapter 8 serious measurements of variable stars are discussed.[101]

Comets

Comets are one of the more interesting and superstition-ridden objects in astronomy. Long believed to be harbingers of disaster signalling the death of monarchs and the loss of battles, we now know they do not affect the lives of Mankind.[102] Most comets circle the Sun just as the planets and asteroids but they tend to have elongated elliptical orbits, stretching far beyond the outer planets and dipping well within the inner Solar system. They also have periods of hundreds to thousands of years so an individual astronomer seldom sees the same comet on two successive returns.[103]

Comets have been best described as dirty snowballs, composed of frozen ices of water and other volatiles mixed with dust and small rocky debris. As they follow their orbits inward toward the Sun, they heat up and the ices evaporate, leaving behind a trail of gasses. Some of the dust also escapes from the comet and forms a hazy ball around the snowball. The pressure of Sunlight and the Solar wind then act on the gas and dust to blow these debris away from the comet. Thus, a tail, always pointing away from the Sun, is formed adjacent to the main body of the comet. As the comet approaches closer to the Sun, the heating becomes more intense and the tail more pronounced.

Most comets appear in the telescope as a fuzzy circular or elongated patch of light and they might be mistaken for galaxies

---

[100] For more information on making visual estimates of star brightness, see the American Association of Variable Star Observers (AAVSO), 25 Birch St., Cambridge, MA 02138, USA. A 20-cm telescope is ideal for estimating the brightness of thousands of stars of interest to the AAVSO.

[101] An excellent reference on variables is *Observing Variable Stars; A Guide for the Beginner*, David H. Levy, Cambridge University Press, 1989.

[102] The one exception occurs when a comet strikes the Earth. A second exception may occur after we venture out into space and start mining them for their materials.

[103] There are a score or more of comets with short periods but they are the exception.

Figure 5.5. Comet Broresen–Metcalf in September, 1989; 4-minute exposure on Kodak 2415 film using a 20-cm Schmidt Camera. Photo by Chris Schur.

but for the fact that they move with respect to the stars. Just a few minutes of careful observation reveals that they change position rapidly. While a score of comets may be discovered each year, only a handful come close enough to the Sun and Earth to show a good tail. Perhaps once a year a comet will become bright enough to be visible without a telescope and once a decade one will be spectacular enough that non-astronomers notice it casually.

The image of a comet is composed of the nucleus, the coma and a tail. The nucleus is a bright condensation at the center of the more diffuse coma. While most casual observers think this is the dirty ice surface of the comet, it is usually not. The nucleus is the most dense concentration of gases and dust near the parent body. Rarely do we see the ice surface itself and then it is usually with very large telescopes, for comet nuclei are, at most, only a few kilometers in size and thus they are intrinsically faint. It is the gas and dust which reflect far more Sunlight than the surface.

The coma is the roughly circular cloud of gas and dust which surrounds the nucleus. Occasionally structures can be seen inside the coma, probably caused by jets of gases spewing off material. The tail, of course, is composed of the gas and dust pushed away from the coma by the Solar wind and Sunlight pressure.

While it may seem that huge quantities of material are required to build a coma a few thousand kilometers in diameter and a tail millions of kilometers long, the gas pressure inside the coma and tail is less than most laboratory vacuum systems can produce. Still, for large comets near the Sun, the rate of casting off material can approach tons per second. Thus, after a few hundred encounters with the Sun, comets will have lost all of their volatiles and appear to be asteroids in highly eccentric orbits.

Comets are classified into two categories based upon their orbits. Comets with measurable elliptical orbits are called periodic comets. Many comets, as they come racing in from beyond the outer planets, have orbits which are virtually indistinguishable from a parabola. In other words, either they will never return again or their periods are in the tens of thousands of years. We simply haven't been able to measure their positions accurately enough to tell the difference.

While there exist lists of periodic comets, most have periods of hundreds of years or orbits which never bring them close to the Sun and thus they do not produce extensive comas or tails. Indeed, there are a few comets which never travel farther inward than Mars and exhibit relatively circular orbits. This makes them look like asteroids except for the fact the large telescopes have detected a small amount of gas and dust escaping from the objects.

Since comets appear unpredictably in the sky, astronomers maintain a communications net in order to notify all observatories when a new one is spotted. The International Astronomical Union (IAU) Central Bureau for Astronomical Telegrams located at the Smithsonian Astrophysical Observatory acts as a central clearing-house for all discoveries. They disseminate the IAU Circulars, notices of discoveries. Once a comet has been reported and verified independently by another astronomer, the IAU sends either electronic mail or postcards to all of its subscribers detailing the position of the comet and its brightness. These data are often posted on several computer bulletin boards. The popular astronomy magazines and astronomy club newsletters often carry listings of comet positions, allowing observers to find them. For most of my years in astronomy, however, I have usually heard about new comets from other astronomers so maintaining contact with one's colleagues is important.

Comets can approach the inner Solar system from any direction – especially those with very long periods. Comets are best viewed when they are relatively close to the Earth within 200 million km or so. They are also brighter as they move closer to the Sun. Once a comet has rounded the Sun it has received its maximum heating and if, on its way back out of the Solar system, it passes near Earth then a maximum show will result.

The comet itself may sport a tail which is so long that it cannot be contained in a single field of view – even with the lowest-power eyepieces. This is a situation which calls for the rich-field adapter as described in Chapter 6 under accessories. Many observers sketch or photograph the appearance of the comet. It can change from hour to hour. Variations in the Solar wind can bend the tail and cause whorls and kinks in it. At times, the tail can appear to disconnect entirely from the coma before re-forming again.

Often it will appear as if bright comets have two tails. One is composed of gas and the other dust. They separate when the forces of the Solar wind and photon pressure are not acting in parallel. Like the coma, the material in the tail is very tenuous and constitutes a fairly good vacuum. Stars can be seen shining through the tail and there is no perceptible dimming of objects behind the tail.

At times, when a comet rounds the Sun, the heating can be great enough to break the comet nucleus into several parts. The most recent example of this in a bright comet occurred to Comet West in 1976. The four parts will slowly drift away from each other but they will have similar orbits, differing from each other by only a slight amount. This means that when they return, they will all arrive within a few years of each other from the same direction in the sky. Astronomers have found several examples of such families of comets.

One method of finding a comet is to go and look for your

own.[104] Many amateur astronomers have discovered comets. There are whole books detailing search strategies, methods of observation and when to look. Generally, hundreds of hours will be spent at the eyepiece before spotting a new comet – or recovering a returning old comet. This is not necessarily a bad thing since in those hours virtually every interesting object in the sky will pop into the eyepiece at one time or another. David Levy, the discoverer or co-discoverer of 16 comets at the time of this writing, gives only three rules for comet hunting.[105] First, observe! He admonishes that a telescope stored with its dust covers on cannot be used to find comets. Second, know what a comet looks like. They can easily be confused with galaxies and nebulae so observing known comets will aid in recognition. Third, know what a comet does not look like. Charles Messier, the great comet discoverer of the late 1700s, created a list of non-comets so that he could rapidly eliminate false comets he had seen earlier. His Messier Catalogue is used by many comet hunters to quickly eliminate known objects.

Most comet hunters search the evening sky in the West and the morning sky in the East. While comets can approach from any direction, they will be brightest when nearer the Sun. This tactic is designed to catch comets as they move out of the glare of the Sun at maximum brightness. It is best to sweep the sky in some systematic manner using the widest field of view eyepiece possible. The rich-field adapter comes in handy here. The sweeping direction can be one of many common search patterns but the observer must ensure that adjacent sweeps overlap so that comets are not lost between fields of view. One observer has removed the wedge from his 20-cm S–C telescope and operates it as an altitude-azimuth mount. He makes long horizontal sweeps parallel to the horizon and then adjusts the elevation (declination control) for the next parallel sweep. Of course, when he finds a comet then he has to record its position by relating it to star charts, for the setting circles on his mount were not meant to be used in this configuration.

Once you think you have found a comet then what should you

---

[104] Once you discover a comet it does indeed become 'your' comet, for comets are named after their discoverers. Occasionally if two or three observers simultaneously and independently discover the same object it is named after all three.

[105] *The Sky: a User's Guide*, David H. Levy, Cambridge University Press, 1991, p. 157.

do about it? First, check the position with respect to the Messier Catalogue and any other references which show the positions of galaxies and nebulae. If it still looks good then sketch its position with respect to the stars in the field of view. This is the base reference to see if the object is moving. At that point you will have a few minutes to check secondary sources of nebulae and galaxies because you're going to have to wait a while before making your next sketch. Wait at least a quarter of an hour before making the next sketch. You might spend some of the time noting in your observing log the size of the coma, whether it shows a central condensation or nucleus and whether it has a tail. If so, how long is it and to which direction does it point?

If, after a quarter of an hour, the object has moved then you probably have a comet. Make another sketch and then wait again. During this wait, compare the position and description of your suspect with the list of known comets in the sky at that time. Is it anywhere near one of the known ones? If so, then congratulate yourself on at least recognizing the object and making an independent discovery – even if you're not the first to see it. At this point if the object is still an unknown, some observers contact a colleague and ask for a confirming observation using a different observing site and telescope. This also serves to check the numbers on the reported location of the comet. Some discoverers have transposed digits in the excitement and haste of a discovery. If you have access to a computer bulletin board then this is a good time to check and see if any new discoveries have been posted from the IAU in the past hour or so.

If, by this point, you are convinced that you have an independent discovery of a new comet then it is time to tell the world via the Central Bureau for Astronomical Telegrams. The discovery will be announced via the IAU Circulars. Their address is given in Appendix 1. They need to know, at a minimum, the location of the comet, the time of observation and an estimate of the direction of motion and its speed. State whether you are working in Epoch 2000 coordinates or some other system. The comet's brightness, size and the presence of a tail should be included. Of course, you will include your own name, address and telephone number. It is helpful for the Bureau to know whether you observed the object visually or photographically.

Having accomplished that, what do you do? Wait for word on

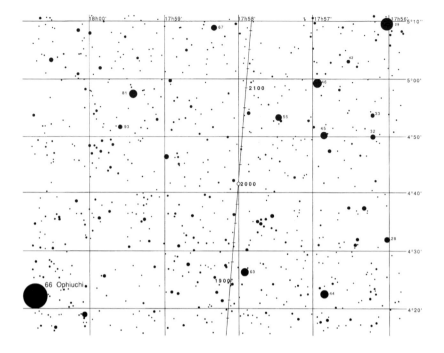

Figure 5.6. Finder chart for Barnard's Star.

your discovery? No, you go back out to the telescope, make one last sketch of the comet and then continue sweeping for your next comet.

The fastest star, Barnard's Star

By now it should be obvious that the Heavens are not constant. Stars explode, comets whirl around the Sun and even the positions of the stars change over the centuries. This is to be expected since our Sun revolves around the center of the Milky Way Galaxy along with all the other visible stars. The inner stars move faster than the outer stars in accordance with the laws of gravitation. We see some stars moving toward us and others going away while still others pass us by in one direction or another. These motions usually take decades to perceive and sensing their movement requires ultra-precise equipment and tedious measurement. There are a few stars, however, which are close so their motion is more readily detectable.

Barnard's Star shows the highest proper motion (velocity at right angles to our line of sight to it) of any star. Located in the constellation Ophiuchus about six light years distant, this Mv 9.5 star appears unimpressive in the eyepiece but do a little experiment. It's located at 17h57'48' and +04°04'02' in Epoch 2000 coordinates, or see Fig. 5.6 for use as a finder chart. Sketch its position with respect to the adjacent stars or take a photograph of it. Then come back in a year and you'll find that it has moved a whopping 10.25 arc seconds North and a little East. In a little under two centuries it will move the Moon's diameter.

This dim, cool, dM5 type red dwarf star has only about 1/2500 the absolute luminosity of the Sun and an apparent visual magnitude of 9.5. With a diameter of 227000 km (140000 miles) it is less than twice the size of Jupiter. It is, however, probably very dense. Part of the reason for its large apparent proper motion is that it is very close. At six light years, only the Alpha/Beta/Proxima Centauri system is closer. In addition, it has an unusually high velocity with respect to the rest of the local group of stars. It moves about 167 km (103 miles) per second; 141 km/s (87 miles/s) of this velocity is toward Earth and 89 km/s (55 miles/s) is at right angles to our line of sight. Due to the large velocity of approach, within eight millennia it will be the closest star to Earth at about four light years. Then it will whiz past us and continue on its journeys. It does not, however, have sufficient velocity to escape the Milky Way Galaxy so it is probably not a visitor from some other galaxy.

If all of the stars moved with this speed then the shape of the constellations would change noticeably within a lifetime. Luckily, the average proper motion of all the naked eye stars is less than a tenth of an arc second per year.[106] Barnard's Star has an added note of interest. As it moves through the sky it does not describe a straight line. Fine positional measurements show it wobbles, indicating the presence of an unseen companion orbiting the star. In actuality, there is evidence from Sproul Observatory that two planets, one with the mass of Jupiter and a second with the mass of Saturn may orbit the star with different periods of revolution.[107]

---

[106] *Astronomy*, seventh edition, 1964, by Robert H. Baker, D. Van Nostrand Company, Inc., p. 324

[107] *Travellers in Space and Time*, Patrick Moore, Doubleday and Company, Inc., 1984, p. 78

The nearest star, Proxima Centauri

This is an exercise for people observing South of latitude 30° North. The closest star to our own Solar system (other than the Sun) is a faint Mv 11 ruddy object which is part of a triple star system. Alpha Centauri is itself a double star located at 14h39'35.885' and -60°50'07.44'. Nearby, at 14h39'36.087' and -60°50'07.14' is the dim dwarf variable companion which orbits the double star Alpha Centauri at a distance of about a sixth of a light year.[108] Like Barnard's Star, the image isn't all that spectacular. Indeed, with a little sloppy observing you might look at the wrong star and never know it. The importance of the star, however, is in knowing that it is the closest. It probably doesn't have planets since stable orbits in a triple star system are rare. Still, it's nice to know who your neighbors are.

Solar observations

Your 20-cm S–C collects enough Solar energy to start fires at the focus. Obviously, looking in the eyepiece at that point is dangerous. You can, however, enjoy many fine hours of observing providing that a proper filter is installed in front of the corrector plate. I'd advise against so-called 'Herschel Wedges' and filters that slip into or over eyepieces. These devices absorb the full concentrated energy of the aperture (about 30 watts of heat energy) and, after being heated and cooled a few times, can crack and suddenly let the full light of the Sun onto your eye.

Filters in front of the corrector either absorb or reflect most of the Solar energy over a wider area than eyepiece filters and thus don't become as hot. The best filters are reflective as opposed to absorptive. Reflective filters are merely a mirror which has been incompletely silvered (modern mirrors use aluminum but by tradition we still call it silvering). Thus, they reflect between 99% and 99.99% of the light. What few photons get through are just about the right intensity for the eye to view comfortably.

However, 20-cm diameter reflective filters are expensive and some people buy sub-diameter 4-cm or 6-cm diameter filters to save costs. Of course, the telescope now has the reduced resolution of a 6-cm telescope which is nowhere near the resolution of a

---

[108] *Travellers in Space and Time*, Patrick Moore, Doubleday and Company, Inc., 1984, p. 74

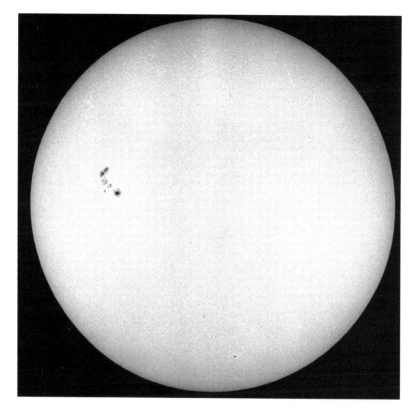

Figure 5.7. The Sun in white light; 1/1000-second exposure at prime focus on Kodak 2415 film. Photo by Richard Hill courtesy of Celestron International[109].

20-cm telescope. On the other hand, in most locations the 2.0 arc second limit of seeing for a 6 cm telescope isn't much of a penalty over the 0.6 arc second seeing for a 20-cm telescope when the average daytime atmospheric limit is about 1.2 arc seconds.[110] I have a full aperture glass Solar filter and I find it only marginally better on most occasions than sub-aperture filters. My home, however, is located in a city notorious for bad daytime seeing.

I'd advise against improvising an aluminized mylar filter in order to save a few pennies. Most aluminized mylar advertized

---

[109] Richard Hill is the Solar Recorder for the Association of Lunar and Planetary Observers (ALPO).

[110] There are places where daytime seeing is much better than this but they are few and far between. They're not hard to locate, however. Just find the nearest professional Solar observatory and set up your telescope in their parking lot.

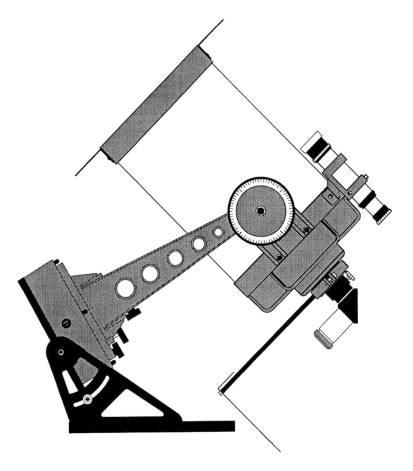

Figure 5.8. Solar observation set-up.

as Solar filter material is designed for direct viewing by the unaided eye. They transmit 1% to 2% of the light to your 0.3 cm diameter eye lens. Your 20-cm diameter telescope, however, has about 4,000 times the collection area of a Sunlight-adapted eye. For a full aperture telescope filter you must allow no more than 0.01% of the incident Solar radiation at all wavelengths to pass.

When attaching a Solar filter, bolt it on. If you have to drill and tap holes in the dew shield lip, then do so. I've seen more than one Solar filter bumped off by careless tourists. Luckily, nobody was looking through the eyepiece at the time. Don't use rubber bands to hold the filter on. They dry out in the heat of the Sun and may let go at any time.

One controversial method of observing the Sun is by projection. In this process, no Solar filter is used and the image of the Sun is directed to a small white card near the eyepiece, as shown in Fig. 5.8. This approach also allows several people to watch the image simultaneously. One caution is that light within the eyepiece is concentrated (though not focused to a point) and eyepiece elements can become rather warm if they aren't of the highest transmission quality. I've heard horror tales of expensive cemented eyepieces cracking under this treatment but I've never been able to track down the rumors to a concrete source. Maybe this is an 'urban legend'[111] in astronomy but I don't use my most expensive eyepiece with Solar projection.

One final caveat on Solar projection; when you point the telescope almost at the Sun, the light is collected by the primary mirror but it winds up falling on the internal baffle tubes which are painted black and thus absorb energy well. While my telescope has metal baffle tubes, if yours are plastic then they can be melted by off-axis Solar light. If you are unsure of the material in your baffles, do not use this method. Additionally, the body of the star diagonal and eyepiece must also be made of metal. I have also been told by one telescope engineer that the heat build-up inside the tube is not good for the components and that stresses can be introduced into the system this way. Although I have observed using this telescope configuration in the past, I shall probably not do it again.

Whatever Solar observing method you use, realize that the dark-colored knobs and fittings of your telescope will become very hot.[112] The fork arms may become too hot to touch. I have made a flat cardboard collar which fits on the front of the telescope and is about 0.4 meters in diameter to shade the rest of the telescope structure. Given a light breeze it shakes the telescope but then again, a light breeze makes the atmospheric seeing go to pot and then it's not worth viewing anyway.

---

[111] An urban legend is an unsubstantiated popular tale, often repeated and of dubious veracity e.g. there are alligators in the New York sewers.

[112] One telescope designer, wishing to solve this problem for a solar-only telescope, had the knobs coated with chrome to reflect Sunlight. The absorption, reflection and radiation properties of chromium in the infrared are such that the fittings become hotter than if they were painted flat black. There exist special white paints which minimize solar heating of surfaces.

As with night-time viewing, I'd advise spending a little time getting comfortable before observing. See if you can arrange some shade for yourself. While I usually sit inside an air-conditioned house and let the telescope and TV camera bake outside, you may not have that option. Be aware of the dangers of sitting out in the Sunlight.[113] It has been said that only mad dogs and Englishmen go out in the Noonday Sun.[114] Add Solar astronomers to that list.[115]

Seeing is always an important consideration but in daylight it becomes more critical. A good sized black asphalt parking lot surrounded by white-roofed buildings can generate upwards of a Megawatt of atmospheric turbulence that you may have to look through. While glider pilots may appreciate such updrafts, they wreak havoc on Solar images. The best daytime seeing is obtained in an environment where the surrounding terrain is thermally homogeneous. An excellent example is Big Bear Solar Observatory, located on a small island in the middle of a lake. I've observed from the middle of wheat fields in Illinois successfully. A trickier approach is to observe at the edge of a precipice with the prevailing wind flowing smoothly (without turbulence) up and over the edge. Mount Wilson's Snow Telescope and Sacramento Peak's Sunspot Observatory are examples of this principle.

Before observing the Sun, make sure the dust cover on the finder scope is firmly in place. The reason is that it, too, will collect the Sun's rays and project them out the back in a nicely focused beam. If you are observing with a filter on the main optics and an eyepiece on a right-angle adapter, then your head will likely fall in the beam coming out of the finder telescope. This beam is concentrated Sunlight and is hot enough to start fires. I have a hat with a burned spot on it to remind me of this problem. In order to line up the telescope with the Sun I look at

---

[113] Shortly after moving to Phoenix, Arizona, and before I became involved in TV astronomy, I spent all afternoon in 45°C (112°F) weather photographing a partial Solar eclipse. The last half hour of my notes and photographs were ruined, as I entered the early stages of heat exhaustion and dehydration, causing lack of concentration on the task at hand.

[114] Noel Coward, popular song.

[115] At international meetings of professional astronomers, it is easy to discern the Solar astronomers from the stellar observers. One group has large pupils and the other has a tan.

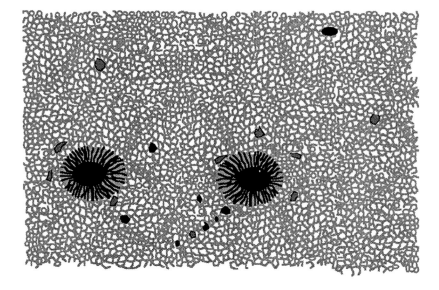

Figure 5.9. Typical Solar features.

the shadow of the tube assembly on the ground and move the tube until its shadow is completely circular.

There are many features to observe on the Sun. The most prominent are Sunspots, areas of slightly cooler temperature on the surface of the Sun. They are still white hot at about 4600K, though – they're just not blindingly white hot as the rest of the Sun is at 5000K–6000K. You may observe the Sun and find no spots, for they form and dissipate over a period of weeks and sometimes there just aren't any visible. Usually spots form in pairs at about the same latitude on the Sun, one preceding the other as the Sun rotates. They will grow in size and separate slightly within a few days. After a week, the spots have usually attained their maximum size and they will then slowly diminish. The following spot generally breaks up and vanishes before the leading spot. Large Sunspots can persist for weeks or even months.

Sunspots are composed of a dark core called the umbra surrounded by a lighter penumbra. While they are generally round, groupings of them can merge into complex shapes. They are associated with magnetic storms on the surface of the Sun and can affect areas several times larger than the Earth. Sunspots are more numerous in some years than others. The number varies with an irregular repeating cycle having a period of about 11 years.[116] At the beginning of each cycle, spots form at about 30° North latitude and 30° South latitude on the Sun. Within two or three years the number of spots increases until there are seldom days when the Sun is free of spots. As the cycle progresses, the spots form closer to the equator.[117] After reaching a maximum count, the number slowly declines until the end of the cycle, at which time the spots form at about 5° North and South latitude. Individual spots, once formed, generally do not move in a North or South direction but they do rotate with the rest of the Sun. One of the activities pursued by many amateurs is recording Sunspots, watching them change day by day. Some observers sketch the images while others use photography. The Association of Lunar and Planetary Observers (ALPO) coordinates observing programs and keeps records of Sunspot numbers, locations and types. Such records are often helpful to professional Solar astronomers who need daily images of the Sun and occasionally miss a day when their own observatories are under clouds.

The majority of the light from the Sun comes from the photosphere, an opaque layer of glowing gas which gives the appearance of a surface. Under conditions of good seeing, granulations or small lighter-colored areas can be seen. These are constantly changing and shifting, usually lasting only a few minutes. They are from 250 km to 1500 km in diameter and are separated by narrow dark spaces. At the limb of the Sun and near some Sunspots bright elongated patches called faculae can be seen.

While the Sun's atmosphere is extremely tenuous above the photosphere, it does absorb some light. For this reason, the limb of the Sun appears darker since light from that region must pass through more atmosphere to reach us. This atmosphere, called the chromosphere, is composed mainly of hydrogen which glows red. Normally it is not seen because of the much brighter glare of the photosphere but there are two methods of observing

---

[116] Since the magnetic polarity of Sunspots reverses at each minimum in the 11-year cycle, some astronomers consider a complete cycle to be 22 years.

[117] The Sun rotates faster at the equator than at the poles, making a rotation in a little less than 25 days at 0° latitude and 33 days at 75° latitude.

it.[118] The first is to travel to a total Solar eclipse, in which the Moon covers the photosphere, allowing the crimson streamers and prominences of the chromosphere to be seen. This can be expensive, the view lasts only a few minutes and Solar eclipses can occur only twice a year at most.

The second method is to use a narrow-bandwidth filter which blocks out all of the light except that of the hydrogen alpha line. Such filters come in several grades, the finer ones having a narrower bandwidth which gives greater contrast in the image. A good hydrogen alpha filter can cost as much as the telescope itself. The filter usually is composed of two filters, a deep red one which fits over the main aperture of the telescope and a smaller narrow band filter which threads onto the back where a camera might normally be mounted. The smaller filter is actually composed of a stack of interference filters and it is very sensitive to temperature. Thus, it usually is equipped with a heater and thermostat control. Changing its temperature actually changes slightly the wavelength of light which it passes. Thus, before each observation the filter must be warmed up and occasionally 'tuned' to the proper wavelength.

The hydrogen alpha filter allows observation of streamers of heated gas which float on magnetic fields within the chromosphere. At the limb of the Sun, the streamers or prominences can be seen extending above the photosphere. At times such displays can form arches and loops, indicating quite a magnetic storm in the Sun's atmosphere.

The presence of violent weather on the Sun reveals some of the underlying mechanism which causes it to shine. The changing magnetic fields which give rise to Sunspots and prominences can also spawn a flare which generates great energy over a small area. Flares are best seen in the light of hydrogen alpha but they have occasionally been observed in white light. They can last from minutes to as long as several hours and occur only a few times per year. In my life I have seen only one and I chanced upon it while testing a new TV camera, as shown in Fig. 5.10.

---

[118] A third method, using an instrument called a coronagraph, is possible but coronagraphs are generally found in large, well endowed professional observatories. While it is possible for an amateur to construct one, I'd recommend obtaining a degree in optics and mechanical engineering first.

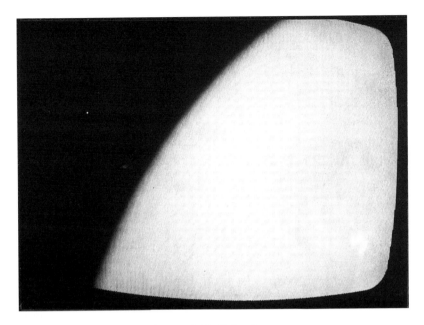

Figure 5.10. Solar flare.

Eclipse chasing

The compact, portable and rugged design of the 20-cm S–C telescope makes it ideally suited to hauling around the world stalking the elusive total Solar eclipse.[119] Participating in the observation of a total Solar eclipse can be an invigorating experience. Often, an eclipse expedition involves travelling half way around the world and setting up in some primitive location for only one or two minutes of totality. It can also be a harried trip in which equipment is set up at the last moment after a wild truck ride along rutted or non-existent roads to avoid clouds. As the shadow of the Moon sweeps across the land at several thousand kilometers per hour you know the eclipse will take place whether you and your equipment are ready or not.

Eclipse chasing is usually a team sport in astronomy. Expeditions of hearty souls pack tons of equipment, passports and tool kits for the adventure. Admittedly, an eclipse is often

[119] At one point in 1980 a Celestron C-8 was the largest telescope in all of Kenya, having been brought there for a Solar eclipse. That distinction lasted for about four minutes until somebody else showed up with a 35.6-cm (14-inch) C-14 telescope.

Figure 5.11. Total Solar eclipse showing the inner corona and promi-nences. February 16, 1980, photo taken at Salt Lick, Kenya; 1/2-second exposure on Kodak Ektachrome 100 film at prime focus. Photo by Bob Fingerhut and Pete Manly.

just the excuse for people who needed a reason to go world trav-elling but valid scientific data are collected on such expeditions by both professionals and amateurs.[120] By measuring the exact time of the beginning and end of totality, information can be gained on the position of the Moon and the exact size of the Sun. In addition, photographs of the corona, when properly docu-mented, help professional astronomers understand the nature of the Sun's outer reaches. This includes the magnetic fields driving

---

[120] Both the International Occultation Timing Association (IOTA) and the Association of Lunar and Planetary Observers (ALPO) have organized expeditions of amateurs for eclipses.

Figure 5.12. Total Solar eclipse equipment being prepared for the 1980 Kenyan eclipse.

the Solar wind. On more than one eclipse, the professionals have been clouded out while amateurs a few kilometers away obtained good results.

Eclipse observation equipment is adapted from general purpose telescopes like the 20-cm S–C. All accessories and observing plans are geared, however, to just a few minutes of time when the Moon obscures the Sun. The telescope in Fig. 5.12 shows some typical modifications. First, the tripod was left at home to save shipping costs and we used the telescope case as a mount, adding a metal plate with the proper fittings at one end. In order to make the mount more stable, the case was filled with sand encased in plastic trash bags. Since the latitude of the observation was about 3.5° below the Equator, the polar axle had to be adjusted until it was nearly horizontal. That caused the telescope tube to strike the case for the particular elevation of the eclipse as seen from Kenya. Thus, a metal plate was added between the wedge and the drive motor base to raise the whole telescope assembly. Finally, since the camera's viewfinder mirror would be locked in the up position to minimize vibrations during expo-

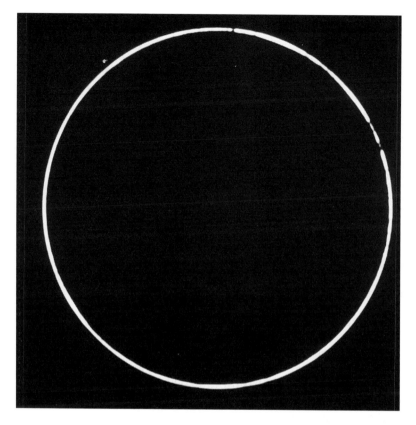

Figure 5.13. Annular Solar eclipse; 1/4-second exposure on
Kodachrome 64 film at prime focus. Photo by Mark Coco, courtesy of
Celestron International.

sures, a precision guiding telescope and tangent coupler were
added.

During the observation, one astronomer guided the telescope
while the other sequenced the camera through a series of exposures of different lengths. Then the astronomers switched positions and repeated the exposure series. While this produced
some nice pictures, the real reason for the trip was to record via
voice tape recorder the exact time that totality began and ended.
Of course, the two weeks spent after the eclipse photographing
wild animals in the game parks wasn't bad, either.

While most eclipse chasers try to attend every total Solar
eclipse, a lesser number of die-hards also chase annular Solar

Figure 5.14. Annular Solar eclipse observing equipment.

eclipses.[121] In an annular eclipse the Sun's photosphere is never completely covered, leaving a ring of light all around the edge of the Moon. This light is bright enough to veil the corona which extends well beyond the surface of the Sun although prominences are often seen during annular eclipses.

Although total Solar eclipses may last several minutes, annular eclipses are seldom more than ten seconds long. Thus, there is a bit more scrambling for data. There are many stories of observers who, in the time pressure of an eclipse, forgot to make some critical adjustment or remove a lens cap and thus travelled half way around the world for no data. With this in mind, I prepared for the May 30, 1984, annular Solar eclipse. The equipment, shown in Fig. 5.14, was designed to run unattended through the several seconds of the eclipse. The main scientific data were taken by the TV camera mounted on top of the main Schmidt–Cassegrain telescope. A small electronic time inserter,

[121] While the Moon and the Sun have nearly the same angular diameter, thus allowing for the geometry of an eclipse to occur, the Moon's orbit is a slight ellipse. At times it is farther away from the Earth and thus has a smaller anguar diameter. When this diameter is less than the Sun's the eclipse is said to be annular since the Moon does not entirely cover the Sun.

Figure 5.15. A very new moon. Photo taken with a 350-mm lens piggy-backed on a 20-cm S–C using a 4-second exposure on Fuji 1600 film. Photo by Bruno Mattern, courtesy of Celestron International.

synchronized to WWV radio time signals, superimposed the exact time on the video signal before it was recorded. The main telescope had a 35-mm film camera at prime focus. This camera had an automatic winder which, after each exposure, advances the film to the next frame. If the exposure button is held down, about two pictures per second are taken. With the video tape recorder running several minutes before the main event, I could disregard its operation during the critical few seconds of annularity. About ten seconds before the Moon was centered on the Sun's disk all I had to do was start the film camera and winder. At some time in the following twenty seconds, it would take a picture of the center of annularity. At that point my main function was to leave the equipment alone and not disturb anything. I could step back and simply enjoy the spectacle.

Seeing very old/new Moons

When the Moon is within a few days of new phase a ghostly silver light often illuminates its dark side. This is caused by

Sunlight striking the Earth first, reflecting off our planet and then lighting up the Moon. When the Moon is nearly at new phase then the Earth is nearly at full phase as viewed by somebody located on the Moon. Imagine standing on a darkened Lunar plain, looking up at a brightly lit Earth, four times larger in diameter than the Moon. Add to that the fact that the Earth generally reflects a greater percentage of Sunlight than Lunar rocks and the scene becomes quite brightly lit. On those occasions when great storms cover the day side of the planet, the white tops of the clouds reflect even more light onto the Moon. Often I have explored the dark side of the Moon with my telescope just to see how much I can see. Since this occurs during twilight, it's usually a time-killing activity reserved for the period after I've finished setting up my telescope and before the sky becomes dark enough to get down to serious observing.

Another twilight activity is searching for very old or very new Moons. When the Moon is 24 hours or less away from the precise instant of new Moon, the Sunlit portions often appear as a single sharp bright line against either a dusk or dawn sky. There appears to be an informal contest as to who can see the Moon closest to the moment of new phase. There are various unofficial categories for completely unassisted naked-eye viewing, binocular or low power finder scope viewing and full telescope observations. Finding a nearly new Moon may take a bit of planning, for not every new Moon is favorable for a given site. From month to month the timing and relative positions of the Sun and Moon change.

First, while the new Moon occurs at that time when the Sun and Moon are at the same right ascension, they may actually be several degrees apart in declination. The farther apart the better since this places the Moon in darker skies. An ideal scenario in the Northern Hemisphere might occur when the Sun is at a far South declination in its trip along the Ecliptic. If the precession of the plane of the Moon's orbit places it at a more Northern declination at the time of new phase then theoretically it might be possible to see the Moon exactly at the time of new phase. This is because the Sun will set at mid latitudes first and be far below the horizon before the Moon, farther North, reaches the horizon. Similarly, at dawn, the Moon would rise before the Sun in this configuration.

In actual practice the situation is a bit more difficult. First, these phenomena occur at very low elevations, seldom more than 5° above the horizon. Thus clear, stable weather and a horizon free of obstacles are mandatory. Second, the Sun is usually just below the horizon and twilight conditions make seeing the Sunlit portion of the crescent difficult. Finally, the plane of the Moon's orbit is inclined with respect to the Ecliptic by only about 5° and thus that is the greatest separation that can occur between the Sun and Moon at new phase. Since the Moon moves along its orbit about 1° every two hours, however, the new Moon becomes easily observable only a few hours from the moment of new phase. Current records for observing both old and new Moons run in the 12–14 hour range but documenting sightings is difficult, although several blurry photographs have been taken.

Plan your initial trials of new Moon sightings for evenings or mornings when the Moon is 24–30 hours away from the moment of new phase. If that moment occurs only four hours before Moonset as seen from your site then you probably aren't going to see it although I won't advise you not to try. The following evening it will be 28 hours old and, weather permitting, you will probably make the observation successfully. My own personal record is about 19 hours and, while no great feat, I did enjoy seeing just how close to new I could see the Moon.

You might ask if there is any practical application for this observation. Indeed there is. Many countries in the Middle East organize their societies around religious calendars which depend on the sighting of the new Moon to indicate the start of the month. While most of these countries simply consult an Astronomical Almanac to determine when the month starts, there are still several nations which rely on visual observations.[122]

Terrestrial observations

More than one telescope has been purchased initially to observe birds or ogle ladies sunning themselves on a nearby beach. There are many uses other than Solar observations for your telescope during the daylight hours. Naturalists have been using them for years to observe and photograph animals from a distance without

---

[122] For a discussion of the techniques and criteria of observing old and young Moons, see *The Astronomical Scrapbook*, Joseph Ashbrook, Sky Publishing, 1984, p. 200.

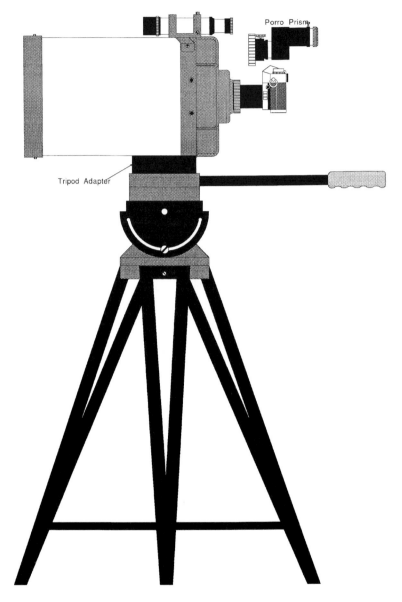

Figure 5.16. Terrestrial observation set-up.

intruding on their environment. A well-known sports magazine uses them to obtain those incredible pictures of the facial expressions on race-car drivers while they're whizzing around the track. At least one aviation writer uses his to capture pictures of racing planes without the danger of standing underneath the raceway.

The equatorial wedge supplied with most telescopes can usually be eliminated and the telescope drive base bolted directly to the tripod. Thus, the telescope can act as a conventional alt/az mount. On some models, as shown in Fig. 5.16, the telescope tube can be removed from the base and fork arms. The manufacturer then supplies, as an optional accessory, a mounting plate with a standard 1/4' × 20 thread which is found on many heavy-duty photographic tripods. This allows the optical assembly to be used like a large telephoto lens.

For daytime photography, I have found that the base and fork assembly which came with the telescope are not capable of holding the telescope steady when the shutter trips. Thus, I use the mounting plate and a heavy-duty tripod designed for the days when TV cameras weighed about 30 kg. On the other hand, I have an old 35-mm camera which goes 'THUNK' when I make a shot. I have seen several people with more modern electronic shutter cameras use the original base and fork arms in an alt/az configuration successfully.

For visual work such as nature studies, the image will appear inverted with most eyepieces. While there are special non-inverting eyepieces available, I'd recommend the use of a Porro Prism such as the one shown in Fig. 5.16. This device goes between the telescope and the eyepiece and uses prisms to invert the image. Thus, all of your normal eyepiece collection can be used. For single-lens-reflex camera photography no image inverter is required since the camera's viewfinder has a built-in prism which inverts the image.

Before using your S–C in daylight you should check to assure that the light baffling is complete. Some early model S–C telescopes allowed sky background light to enter the telescope at a point adjacent to the secondary mirror at an angle off the optical axis. Such a ray would not strike the primary mirror but would go down the Cassegrain hole in the primary mirror. Since the sky background is mostly black, this unfocused light wasn't thought to be a serious problem. It is a problem when viewing

Lunar or Solar features. The central baffle tube is designed to block out such rays but there are some baffle tubes which are too short.

In order to see if your telescope has this problem, point the scope at a bright diffuse source like the daylight sky (but not directly at the Sun). Remove the eyepiece and look down the baffle tube. You should be able to see only the secondary mirror and the image of the primary mirror reflected in the secondary mirror. If you can see any light outside of the secondary mirror support, as shown in Fig. A5, then the optical system is inadequately baffled. Move your eye from side to side and peer up the baffle tube from the edge of the hole the eyepiece slips into.[123] The usual fix for stray background light is to make a black cardboard annulus which fits around the secondary mirror. While this will decrease the effective aperture (light-collecting area) of the optical system it will probably not be a significant decrease.

Your telescope can be used with incomplete baffling but any bright object just outside the field of view may produce a hazy brightening of the entire field, thus losing the fine details of low contrast image features. While this usually isn't much of a problem at night, daylight use implies that there will almost always be bright objects just outside the field of view.

Artificial Earth satellites

Most of us think of satellites as points of light moving rapidly through the sky, best seen at dusk and dawn. Their motions, which may approach one degree per second, are much too fast to follow with a 20-cm S–C on commercial mountings. Further, they seldom move only in RA or only in declination. Thus, complex two-axis tracking would be required.[124] The optics from 20-cm S–C telescopes has been attached to special satellite tracking mounts but then the S–C usually becomes the finder telescope for acquiring satellites and a much larger main telescope is used to collect the data.

---

[123] I have seen one S–C incorporating a custom TV camera adapter with a mounting hole which is larger than usual. While marginal rays can get down the baffle tube and out the back of the telescope, the special camera itself has a slightly smaller sensitive area and thus background rays cannot appear in the TV picture.

[124] The paths of satellites are complex but they may be approximated as great circles centered on the observer. A three-axis mount can be designed to track satellites by driving a single axis but the drive speed must have a wide dynamic range.

Several amateur astronomers told me that they have released both axle clutches and hand-guided their telescopes while visually observing either the Mir space station or the Space Shuttle. While the image of such large spacecraft would theoretically be several arc seconds in size when at a range of 200 km to 500 km, the hand-guidance required is extremely difficult. Generally, acquisition takes two people, one looking through the finder scope and guiding until the spacecraft appears in the main telescope where the other person takes over control. The hand-off is a very tricky maneuver and requires considerable practice. I tried a similar observation in 1977 on the Skylab spacecraft a few days before it re-entered the atmosphere. With suitable orbit data from NASA, it was not difficult to acquire the spacecraft in the finder scope. I couldn't place it steadily in the main telescope. When I looked through the main scope and let my observing companion use the finder scope, he never had a steady enough view to hand off control of the telescope to me. I saw the spacecraft streak through the main telescope's field of view a couple of times but the motion of the telescope was so fast and erratic that I could see only a flash with no detail.

There is, however, a class of satellites which move only in right ascension and they move slowly enough that they can be tracked steadily. These are the synchronous satellites, almost 36,000 km from the Earth's surface. While their distance makes them faint objects to the naked eye, some of them are as large as a pickup truck and thus they appear as Mv 8 to Mv 12 starlike objects. Several communications satellites use large parabolic antennas which were stowed during launch and unfolded after placement on orbit. These satellites may appear as bright as Mv 6.

The synchronous satellites, sometimes called syncsats, have an orbital period of 24 hours and thus they do not move with respect to the Earth. They are designed to remain above one spot on the Earth forever.[125] Thus, they move with respect to the stars at a rate of 15 arc seconds per second Eastward.

---

[125] The Solar wind and small gravity gradients will nudge these spacecraft into different orbits over a period of months or years. Spacecraft owners, who want the satellite to stay put, use small station-keeping rocket motors to periodically adjust the position of their satellites. Obsolete or malfunctioning satellites are usually moved out of the crowded synchronous belt.

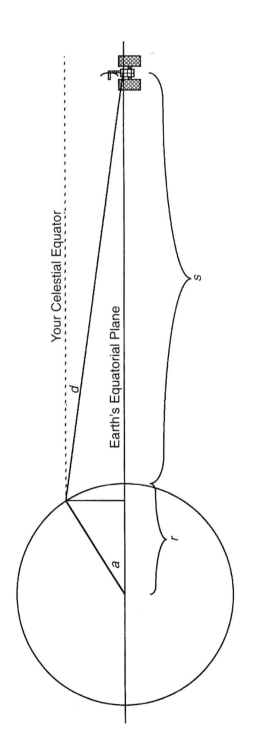

Figure 5.17. Synchronous satellite positions.

All of the satellites orbit in the plane of the Earth's equator.[126] The place to look for them, however, is not on the Celestial Equator. The satellites are on the plane of the Earth's Equator which is parallel to the Celestial Equator but separated from it by an angle which depends on your latitude. The diagram in Fig. 5.17 shows the relationship between your latitude, a, and the declination, d, which you must search to find synchronous satellites, where $r = 6,378$ km, the Earth's Equatorial radius and $s = 35,807$ km, the altitude of synchronous satellites. A general solution to that trigonometry problem is;

$$d = \tan\text{-}1 \left[r \sin a \ / \ [s + r - [r \cos a]] \right]$$

Evaluation of that equation for various latitudes is shown in the table below. Note that Northern Hemisphere observers should search at negative declinations while Southern Hemisphere observers should search at positive declinations.

Latitude	Declination
0°	0°
10°	-1.8°
20°	-3.4°
30°	-5.0°
40°	-6.3°
50°	-7.3°
60°	-8.1°
70°	-8.5°
80°	-8.7°

In order to find synchronous satellites visually, I use a wide-field-of-view eyepiece. Set your declination to the proper value and search an hour of right ascension or so on either side of the anti-Solar point. In other words, early in the evening look in the East while just before dawn, look in the West. This assures that the spacecraft you observe are illuminated 'face on' by the Sun. While you can observe spacecraft farther away from the anti-

---

[126] Some synchronous satellites have an orbital plane which is tilted slightly with respect to the Earth's equatorial plane. Thus, they will appear to nod North and South a degree or two with a period of one day. Similarly, if their orbit is slightly elliptical instead of the usual circular orbit, they will appear to drift a degree or two East/West, returning to any given starting point in 24 hours. The combination of these two motions often makes the spacecraft trace out a figure 8 in the sky.

Solar point, they will be illuminated from the side or possibly slightly back-lit and thus will appear fainter.

There are a hundred or more objects in the synchronous belt counting some spent second- or third-stage rockets, several inoperative spacecraft and disposable artifacts such as aerodynamic shrouds and optical window covers. I've never had to spend more than about five minutes searching for one. Several times I have had two or more in a wide-field eyepiece simultaneously. Many of the satellites rotate in order to stabilize themselves like a gyroscope and Sunlight glinting off their sides will produce flashes and fades as different parts of the spacecraft are illuminated.

They appear visually like stars which drift Eastward through the background of real star fields. Usually they will be visible all night long. The exception is a couple of weeks when the Sun is near one of the Equinoxes. During that season, each synchronous satellite will spend an hour or so in the Earth's shadow. Just before going into Earth shadow some satellites will produce huge flashes of light to Mv 2. My guess is that they are stabilized pointing at Earth and there is a large flat shiny surface on the front, probably composed of thermal reflecting foil seen on many spacecraft. Just before going into eclipse, the Sun-spacecraft Earth angle is at a minimum and specular reflections are most likely.

Tracking synchronous satellites is easy. Since they don't move with respect to the Earth, turning your drive system off will assure tracking.

How far can you see?

The slick answer is that when you look between the stars, you're actually watching the 3K background radiation left over from the Big Bang. That's not very satisfying since it looks just like the background sky glow caused by Moonlight or the lights of a nearby city.[127] There is a real question, however, of how far you can see with a 20-cm aperture. Well, it depends on the brightness of the source, of course.

---

[127] For a cloudy night observation, one can see the 3K background by turning on a TV and tuning it to a blank channel. If the contrast is turned up and the brightness down, just a series of flickering bright spots is left. Most are receiver noise but one or two percent of the spots are from the microwave background.

The brightest objects in the Universe aren't stars or even giant elliptical galaxies. They are quasars or quasi-stellar objects (originally called quasi-stellar radio sources). They initially came to astronomers' attention as radio sources. When astronomers found the optical counterparts they appeared like faint blue stars. The puzzle deepened when their spectra showed broad emission lines at wavelengths which didn't correspond to light emission phenomena in familiar stars and galaxies. In 1963 Maarten Schmidt finally unraveled the mystery when he recognized that the spectrum looked like normal hydrogen emission if one assumed that the object was rushing away from us at more than one tenth the speed of light.[128] The object obviously wasn't from nearby, for its speed of 44,000 km/s exceeds the escape velocity for our Galaxy.

In order for this strange object, brightest of the quasars at Mv 12, to be marching in order with the rest of the Universe, it must obey Hubble's Law which states that the farther an object is from us, the faster it must travel away from us.[129] This is due to the general expansion of the Universe and it implies that 3C273;[130] the closest quasar, must be about two billion light years distant.[131] Since we know its apparent brightness and its distance, we can calculate its absolute brightness, yielding the staggering output of six trillion Suns. Not even galaxies come this large.[132] Add in the intriguing fact that some quasars vary in brightness over a period of weeks and the mystery just deepens, for whatever mechanism causes the variation, its physical size must be on the

---

[128] While this was a preposterous concept in those days, it has generally been accepted. For the fascinating story of how this discovery was made, see *Black Holes, Quasars and the Universe*, second edition, Harry L. Shipman, Houghton Mifflin Co., 1980.

[129] There is still debate on the Hubble parameter but current best estimates are that it lies between 50 and 100 kilometers per second per megaparsec. I have used a value of 75 kilometers per second per megaparsec.

[130] The designation 3C273 indicates that this is the 273rd object found in the third Cambridge radio survey.

[131] At the time of discovery, considerable discussion centered on the possibility that quasars appeared artificially distant and really resided much closer to home. Such theories either postulated a non-Doppler shift mechanism to account for the spectrum or discounted the spectral interpretation. Even today there is some small debate on the subject. As an example, see the discussion in *Sky & Telescope Magazine*, July 1990, p. 7.

[132] Recent observations of 3C271 by J.B. Hutchings and S. G. Neff have revealed that it resides within a host galaxy and produces a jet of material flowing away from it. See the *Publications of the Astronomical Society of the Pacific*, Volume 103, Number 659, p. 26.

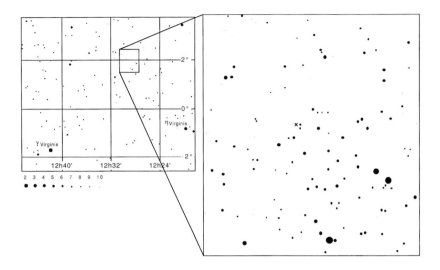

Figure 5.18. Finder chart for quasar 3C273.

order of just a few light weeks of travel at the speed of light.[133]
Astronomers were, and still are, debating how one gets the output of six trillion Suns from such a small volume. The obvious
candidate is a massive black hole but researchers are still working on the details of how the object shines so brightly.

You can find 3C273 in the constellation of Virgo. Its coordinates are listed in Appendix 8 and the finder chart in Fig. 5.18
shows the location of this Mv 12.8 object. While it appears as a
rather unimpressive speck of light, I found it interesting to contemplate that the photons I viewed had travelled for about two
billion years to reach my eye. Such images are sometimes
referred to as fossil light.[134] The photons from 3C273 started their
journey before our ancestors had accomplished the feat of crawling out of the seas onto the land. Think about that a little.

---

[133] This presumes, of course, that whatever mechanism changes the brightness of the
quasar affects the entire quasar.

[134] For more on 3C273 see *Journal of the Astronomical Society of the Pacific*, January 1991, p.
26.

6

Some accessories for the telescope

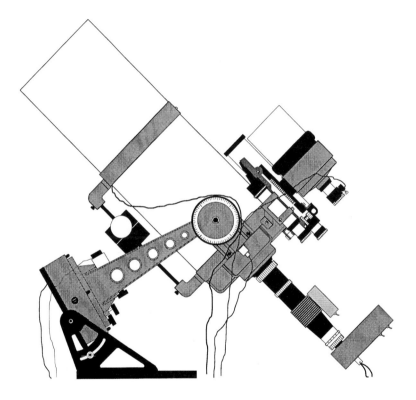

Figure 6.1. A loaded telescope with two finder telescopes, tangent coupler with 90-mm aperture Maksutov guiding telescope and illuminated reticle eyepiece, intensified CCD camera, dew shield and heater, sliding bar counterweight set, electric focus motor and electronic drive corrector.[135]

---

[135] Tom Johnson, one of the original designers of the C-8 telescope, took a look at this set-up and remarked that he'd never thought the telescope bearings would have to carry as much weight as I'd piled on.

Most accessories from one telescope manufacturer will fit on other brands, although counterweights are a notable exception. There are basically two systems of optical accessories; the ones that screw on to the thread on the back of a Meade/Celestron and the T-system, historically developed for the photographic camera industry. Adapters are available for changing between the two systems and coupling to cameras.

Changing the effective focal length

Once you have tried photography at the prime focus, you might find some astronomical images a bit small on the negative. The first approach to this problem is to purchase a tele-extender.[136] This is a small negative lens which fits just in front of the camera and effectively doubles the focal length of the telescope. Since the camera sees a light beam from an f/20 telescope, exposures of extended objects will have to be about four times as long as at prime focus. The longer focal length also means a smaller field of view and thus precise focusing and atmospheric seeing become more critical. For longer exposures, the quality of guiding must also be higher or the star images will be trailed. For even higher magnifications, an eyepiece may be inserted into the light path to act as a magnifier. This is described in Chapter 9.

For larger astronomical objects such as the Orion Nebula, a wider field of view is desired and so a tele-compressor can be purchased. This is a small positive lens which mounts just in front of the camera. I have two of them. One effectively halves the focal length of the telescope, making it an f/5 system. Exposures need to be only a quarter of what they are at prime focus. The lens, however, has a problem in that the image at the edge of the frame isn't as bright as at the center. This is called vignetting or portholing. For most photos where the corners of the picture are black sky anyway, the effect isn't noticeable but if one were to photometrically measure the brightness of the edge stars, they would prove to be noticeably fainter. The second type of tele-compressor reduces the effective f number

---

[136] Unfortunately, the term "tele-extender" has been applied to two different accessories with similar purposes but very different designs. The eyepiece projection tube as shown in Fig. 9.2 has also been called a tele-extender.

to f/6.5 which yields a wide field of view without the dark corners.[137]

A related accessory is the rich-field adapter, or visual reducer, a set of fittings which allows a tele-compressor to be coupled to an eyepiece, resulting in wider fields of view. Since the tele-compressor decreases the back focal distance[138] of the optical system, the eyepiece must be coupled closely to the telescope.

For the larger deep sky objects this combination often results in superb views. The effective f number of the telescope is, however, cut in half and thus the background sky brightness is quadrupled. I find the rich-field adapter to be useful only under the darkest skies. On the other hand, nebulae also appear four times brighter, if only half their regular size. This accessory is also useful for searching for comets when their exact position is a little uncertain. The wider field of view allows more sky to be searched quickly.

Binocular eyepiece adapter

The binocular eyepiece adapter allows both eyes to be used simultaneously, which is a more natural way of viewing for most people. The image brightness is, of course, cut in half for each eye but for bright objects this does not matter. For faint objects and stars, however, detail can be lost if the brightness level of some feature as seen in a single eyepiece is at or near the threshold of detection for the eye. When that level is cut in half, the brightness of the image goes below the threshold and nothing is seen.

For some galaxies, one gets a three-dimensional impression of the nebula hanging in space. Of course, since both eyes are receiving exactly the same image, this is not true stereoscopic vision, but an optical illusion. It is, however, a very pleasant illusion. Some observers maintain that they can see more detail

---

[137] The discussion of tele-extenders and tele-compressors relates to photography with the standard 35-mm format. When used with smaller-format sensors such as CCD cameras, more-powerful tele-compressors have been used with focal length reductions of up to a factor of six.

[138] The back focal distance is the length between the mounting plate on the back of the telescope and the image plane for an object at infinity. With the focus knob adjusted at its extreme, most 20-cm S–C systems allow several centimeters of spacing but a tele-compressor will decrease this by a factor of two.

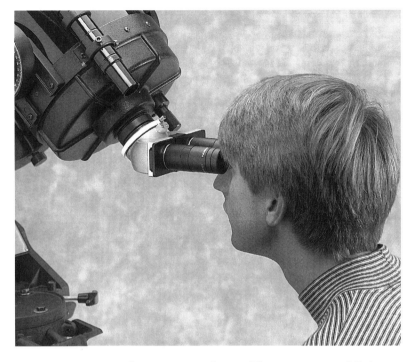

Figure 6.2. Binocular eyepiece adapter. Photo courtesy of Orion
Telescope Center, used with permission.

with this device than when using only one eye but, theoretically,
the single eyepiece should be better since it has fewer optical components which might scatter light. The binocular eyepiece adapter
introduces several prisms and mirrors into the optical train.

One disadvantage is that you have to buy twice as many eyepieces. Both eyepieces should be identical in manufacturer and
model number, preferably bought from the same lot. I have tried
to use two seemingly identical eyepieces of the same focal length
and field of view but made by different manufacturers. Although
the specifications were the same, the differences were great
enough to give me a headache after only a few minutes of
observing.

Making things handy

I am a firm believer in maintaining personal comfort while

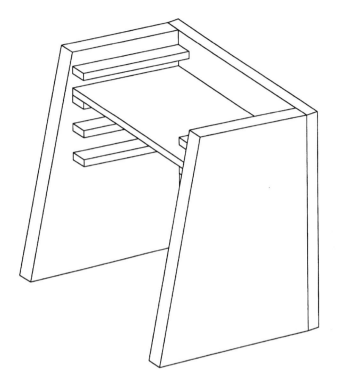

Figure 6.3. Adjustable-height observer's chair.

an adjustable-height observing chair.[139] This allowed me to sit comfortably, no matter what the elevation of the eyepiece. For long observing sessions, this enabled me to concentrate on astronomy and not on an aching back caused by bending over to accomodate some awkward eyepiece orientation.

A second handy accessory is a metal tray which fits between the legs of the tripod and has fitted holes for eyepieces and a small working surface for maps, flashlights, etc. I've never used one of these since by the time they were invented, my normal compliment of required accessories and junk far exceeded the capacity of the tray. Thus, I have a small table hooked to the side of my van which holds these items.[140] The important point here is

---

[139] I now observe from inside the house, leaving the remotely controlled telescope and CCD camera out in the cold.
[140] For note taking I have a small clip board made for pilots which straps to my thigh and is thus always handy. The board has an attached red light.

that eyepieces, maps and such should always be within arm's reach. If your accessory table has become so large that you cannot reach all the way across it then you have too many accessories.

One of the nicer accessories for telescopes without a permanent dome is a wheeled dolly which fits on the ends of the tripod legs. One of the schools nearby leaves their telescopes set up on these wheels. When students want to use the telescope it is rolled out into a courtyard and set up. The dolly framework has jacks which raise the wheels off the ground to improve stability. The instructor intends one of these days to align the telescope carefully on the pole and then mark the jack positions so he can install locating sockets for the jack pins. This will decrease the time required for telescope set-up. While the newer models of commercially made tripods allow for adjustment of the height of the wedge, I have seen a few installations where the observer has rebuilt the tripod so that it is at a convenient height for observers in wheelchairs. I have also seen a platform built around the tripod to accomodate wheelchairs but this appears to be a more cumbersome solution than modifying the tripod.

Many observers have installed a permanent pier in the back yard which is carefully aligned to the Pole Star. When the observer wants to use the telescope, only the tube assembly and base with the drive motor need to be moved, thus speeding the set-up time required. Plans for such piers are often included in the telescope manual. Usually, they are made from 10 cm or larger diameter water pipe and sunk into a concrete foundation. A typical installation is shown in Fig. 10.3. One word of caution; I have seen permanent piers assembled from scrap angle iron welded into a tripod shape. These generally are less stable than thick-walled water pipe and some vibrate like a bell when tapped on the side, resulting in a smeared image for several seconds until the vibrations dampen out.

If you ever intend to build an observatory around your permanent pier then one step is necessary before installing the mount. When the observatory is built, its floor and walls should be isolated mechanically from the pier so that people walking on the floor do not vibrate the telescope.[141] Typically, the observatory

---

[141] For a complete discussion on observatory structural design, see *How to Build Your Own Observatory*, third edition, Richard Berry, Kalmbach Publishing Company, 1991.

floor is built as a bridge from one side of the foundation to the other while not touching the pier or the ground near the pier. In order to further isolate the telescope from the building, a layer of sand at least 30 cm thick should completely surround the concrete into which the pier is sunk.[142] It is much easier to install the sand initially than to dig up the pier just before building the observatory.

Finder telescopes

The six-power 30-mm aperture finder telescope found on most 20-cm S–C telescopes is usually good enough to locate all of the Messier objects and planets except Pluto. Since I often have to locate calibration stars which tend to be on the faint side, I have added an eight-power 50-mm aperture finder telescope. This is probably overkill for casual stargazing but it has come in handy at public star parties when the sky has been less than ideal and I've been frustrated in finding some fainter objects. Prices on such finder telescopes are, in my opinion, excessive and I wouldn't have bought mine if I hadn't found it at an astronomical swap-meet.

In Chapter 2 the problem of straight-through versus right-angle finder scopes was covered. Both styles have their advantages and disadvantages. I've solved the problem by mounting one of each type. I didn't design that unique solution – when I bought the second finder it just happened to be the type complimenting the original finder. Fig. 10.10 shows the telescope with a 'super finder' which is a 90-mm aperture Maksutov telescope with an 80-power eyepiece. This is a specialized finder used when working with very small field of view detectors, as described in Chapter 10.[143]

There is a class of finders which are not telescopes at all, but rather one-power aiming devices similar to the front and rear sights on a rifle. For initial pointing I quite often sight along the alignment screw heads on my larger finder while other people just look along the side of the telescope tube (I can't – I've got too

[142] Some observers have further increased the stability of a pier made from water pipe by filling the pipe with sand or oil.

[143] I have seen one installation in which a 20-cm S–C telescope was used as a "super finder" for a larger, half-meter class photometric telescope.

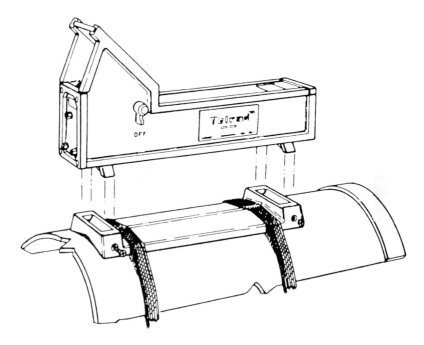

Figure 6.4. Telrad reflex sight. Illustration courtesy of Steve Kufeld, Telrad Incorporated.

many accessories in the way). This presumes, however, that there is enough light to see your telescope and its fixtures.

Several designers have come up with illuminated one power projection systems or reflex finders, as shown in Fig. 6.4. The observer looks at the sky through a partially reflecting glass which has a cross-hair or bull's-eye superimposed on the sky image. In most systems, the image of the cross-hair is focused at infinity so that when you look through it, the stars and the aiming device are both in focus. This finder is most useful for those who practice the star hopping method of finding objects, as discussed in Chapter 2.

Flashlights

I have a shoe box filled to overflowing with small flashlights which I have discarded. Some have clever designs such as the one which clips to my lapel so that I can work with both hands free. Unfortunately, I seldom observe in a business suit which is

144

the only clothing I own with lapels. One has a magnet and swivel head so it may be secured anywhere and its beam directed as needed. Alas, the entire telescope and all the accessories are either aluminum, plastic or glass which are non-magnetic. Many of the discards are small penlights of various designs, most of which have succumbed to the Drool Effect, for I tend to hold the flashlight in my teeth when working on the telescope and eventually the electrical parts of the flashlight become rusted.

I have found a penlight which is sealed with O-rings and can be used underwater.[144] It cost $15.00, ten times the normal price for a penlight but then again, it has lasted for seven years and none of the others made it through six months of observing. It is much too bright for astronomical use so I have fashioned a small red cap which fits over the end and dims the light to the proper level. The cap must be red, for that color preserves night vision, as described in Chapter 3. Finally, the flashlight has a small loop for a lanyard to hang around my neck so I can find it in the dark.

Recently several manufacturers have produced flashlights using light emitting diodes (LEDs) which emit a deep red light and are much more efficient than incandescent white light bulbs. Some clever manufacturers have built LED lights for reading star charts into the hand controller for the drive corrector. I am waiting for somebody to produce an LED penlight which can withstand the Drool Effect.

Drive correctors

Most fork mounted S–C telescopes have a built-in drive motor which approximately matches the rotation of the Earth. This is powered with conventional 115 or 220 V supply commonly used for electrical appliances. The alternating current runs at either 50 Hz or 60 Hz, depending on the country, and drives a synchronous motor.[145] Synchronous motors stay exactly in step with the alternating current of the power source. Thus, their speed is governed by the 50 Hz or 60 Hz frequency. While power companies maintain the accuracy of the frequency to high standards, two problems emerge. First, the stars don't move at a speed which is

---

[144] It is also handy for stirring coffee or hot chocolate on cold observing nights.
[145] For specialized applications, both stepping motors and servo-motors have been used but these are rare.

a simple fraction of the power company frequency (but the Sun does move at that speed). Second, between the motor and the telescope are a series of gears. Gears, almost by definition, have slight imperfections. While a gear will maintain a given speed ratio with its mating gear over an entire revolution, as each tooth engages slight differences in rotation speed will result. This is called periodic gear error.

If you are observing a star visually at high power this effect may appear in the eyepiece as a slight wavering back and forth over a period of a minute or more. It is not a serious problem but for astrophotographers it can smear the image hopelessly. The solution is a drive corrector. This device takes the standard power frequency and changes it slightly.[146] Drive correctors have a control which allows the observer to change the speed of the motor by a few percent, thus accommodating gear problems. Some correctors also include separate speed settings for Solar and Lunar drive rates.[147]

One might ask how much accuracy in drive speed is needed. If you are going to observe the Moon or planets at low magnification visually, then just plug the telescope into a wall outlet and every several minutes you can adjust the pointing. You don't need a drive corrector. If you're going to observe in the field where there is no power, then perhaps a drive corrector would be handy, for most of them optionally allow power input from a standard 12 volt automobile battery.[148]

If you are going to engage in astrophotography then a drive corrector is certainly needed, as discussed in Chapter 9. Many of the models offered also have provision for controlling an optional declination motor and illuminating an eyepiece reticle.

---

[146] The typical drive corrector first converts the power line voltage using a transformer to 12 volt alternating current . The voltage is then rectified into direct current and a variable frequency generator produces 12 volt alternating current. This is then fed to another transformer which increases the voltage to standard power line levels.

[147] One manufacturer's computerized drive corrector measures the observer's guiding commands over one revolution of the final drive gear and then applies those corrections for subsequent rotations.

[148] Most drive correctors are high-precision electronic devices capable of performing under the rigorous demands of astrophotography. They cost $100.00 and up. If you are sure that you will observe only visually, you can purchase a voltage inverter made for operating small appliances at camping supply stores for about $30.00. This will match the drive speed well enough for visual work. Be careful that the rated wattage of the inverter is greater than the power consumed by your drive motors.

One even has provision for an anti-dew heater. Some of them boast incredible accuracy in frequency control. Alas, using a one part per billion accurate drive corrector on a telescope which has a spur gear error of one part in a hundred is a wasted precision.

Within the past few years manufacturers have started producing 20-cm S–C telescopes with a worm and wheel gear drive instead of the older spur gear drive. The worm drive design has much less periodic gear error but a small residual does remain as long as spur gears are used anywhere in the drive train. For most visual observers the newer design seems to eliminate periodic gear error problems but photographers report that they must still guide astrophotos. The amount of correction required is less and the reaction of the telescope to control commands is more positive with the newer design.

Focus motors

I'd always considered this one of the more frivolous accessories until I started working with small intensified TV cameras on my S–C. The usual method of focusing is to monitor the video signal of a particular pixel on which a star is being focused. Unfortunately, merely touching the telescope's focus knob causes the star image to wander several pixels away. The focus motor eliminates this problem. It has also been handy when focusing photographic cameras and photometers. I do believe I achieve a better focus with my film camera using a focus motor. Most focus motors use a simple d.c. motor with an in/off/out switch. The better ones have a speed control. Mine is a bit more sophisticated, using a custom-designed stepping motor, gear train and counter/computer interface so that I can return precisely to any previously known focus setting. Its complexity and associated maintenance over the years may indicate, however, that it is an overdesigned device.

Dew control

One of the main defects of the S–C as cited by Newtonian owners is its susceptibility to dew. Condensation will occur when the temperature of the corrector plate is at or below the dew point, which is a function of the air temperature and the local humidity.

If the air temperature falls below this point then water will condense spontaneously and you have rain. The corrector plate can be significantly colder than the surrounding air because the corrector is radiating heat in all directions to the night sky, which has a very cold effective temperature. In a Newtonian, while the mirror is radiating to the sky, the slightly warmer inner-tube surface is radiating heat to the mirror. The cold sky, looked at from the point of view of a Newtonian mirror, is only a small area at the distant end of the warm tube which fills most of its hemispherical thermal radiation field of view.

As with most observing problems, it is better to prevent the dew from forming in the first place rather than try to get rid of it after it forms. You can prevent (or at least forestall) dew with a dew cap or dew shield. This is an extension of the telescope tube beyond the corrector plate. While this makes the telescope longer and thus more susceptible to wind gusts, the dew shield need not be bulky. Many people use a brown paper grocery shopping bag with the bottom cut out. At one time I used a thin black foam rubber sheet rolled into a cylinder. This hi-tech approach, in which the foam continued down the tube past the declination attachment points, was intended to thermally stabilize the entire telescope. Alas, it worked about as well as the brown bag approach. My current dew shield, shown in Fig. 6.5, is actually a second whole telescope tube, obtained from the telescope manufacturer's scrap heap of slightly defective parts. It is painted black on the inside and white on the outside and keeps the dew away longer than any other shield I've tried. Of course, when operating in the field it's too bulky to drag along so I use a brown bag. One final dew shield is an empty ice-cream bucket. A famous chain of ice cream stores which features 31 flavors ships the confection in five-gallon buckets. When the bottom of the bucket is knocked out, the diameter just happens to be the same as the front end of a 20-cm S–C. Store managers are usually happy to give away empty buckets or you may wish to buy a full one and empty it yourself. I recommend chocolate fudge ripple for best observing results.

While dew shields extending the tube out beyond the corrector plate are a passive approach to keeping the corrector plate warm, there will be some nights with particularly high humidity and an unusually cold sky. Under these conditions a small anti-

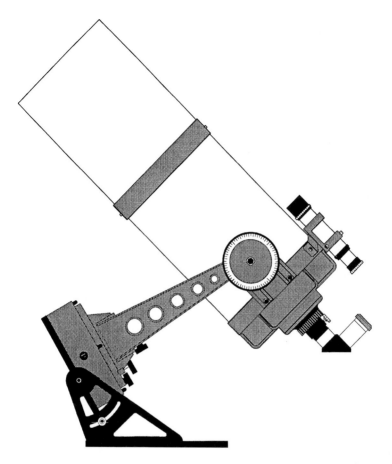

Figure 6.5. Dew shield.

dew heater installed around the front end of the telescope is required. Less than 10 watts of electrical heat are required for even the heaviest dews. The heater is either a string of resistors or multiple turns of nichrome heater wire. The total resistance depends on the voltage source you intend using.

The theory is to keep the corrector plate just slightly warmer than the dew point. While many people take the 'more is better' approach, use of excessive heat will probably cause air currents which can inhibit seeing. In extreme cases, overheating the corrector plate may cause thermal stresses which could cause it to crack.

In the event that dew forms I do not recommend wiping it off with a lens cleaning tissue. You might cause a small piece of dust

149

on the plate to scratch the surface as you wipe. Wiping doesn't warm the corrector plate and within minutes more dew will form.[149] The proper approach is to warm the corrector using the above techniques or apply a stream of warm air (gently). A hair drier gun on one of the lower heat settings or similar devices designed to remove ice from automobile windshields may be used. This is strong medicine and while I've never heard of anybody thermally shocking a corrector plate so much that it breaks, I'd be wary.

The primary mirror and the inside surface of the corrector plate in an S–C will almost never dew up unless you live in a tropical rain forest. If they do, the situation is just about hopeless until you can warm the entire tube and drive the water vapor out. This will take many hours. You can, however, take preventative measures. The only way condensation can occur within the telescope is to first fill it with moist air and then cool the entire telescope assembly. This can happen with telescopes left permanently outside or in an unheated dome. The secret is to keep the inside of the tube filled with dry air. Since not much air flows into or out of the tube when changing eyepieces, it's not a serious problem during observations.

I have modified the rear dust cap of my telescope, however, by taping a packet of water-absorbent silica-gel crystals to the inside of the cap. Thus, while the telescope isn't being used it's being dried out. When the crystals become saturated with water they can be dried out in an oven and reused.

Electronic setting circles

Electronic setting circles attach to the axles of the telescope and measure its position electronically.[150] For most models, the

---

[149] I once wiped the dew off when I was moments from an important asteroid occultation, which would occur whether or not I was ready to observe. I also knew that I wouldn't have to observe for more than about five minutes until the occultation would be over. This is the exception that proves the rule. I risked scratching the corrector to get the data. After that incident I paid more attention to dew prevention.

[150] There are two types of rotary encoders; absolute and incremental. Absolute encoders do not need calibration. Incremental encoders merely measure the direction, amount and speed of the axle rotation, relying on a small microprocessor to sum the motions and keep track of the absolute pointing angles of the scope. Accurate absolute encoders are very expensive so most electronic setting circles use the incremental type.

observer must initially tell the encoders where the telescope is pointing at the beginning of the night. The device then keeps track of the sum of all axle motion and displays the right ascension and declination of any other place the telescope might be pointed at.

They seemed to me to be a frivolous accessory when I first saw them. Then one partly cloudy night I was trying to line up on a faint star for an asteroid occultation. I could see only small patches of sky and few key bright stars. Fortunately, a friend was demonstrating his new electronic setting circles and we were able to calibrate them on a couple of bright stars before zeroing in on the correct star for the occultation. As it turns out, clouds closed in later but I can see how they make finding particular coordinates in the sky much easier, especially under a bit of pressure such as locating the star before the occultation occurs. On the other hand, their utility hasn't yet caused me to reach into my bank account and buy a set.

Some electronic setting circles include a small microprocessor, as discussed in Chapter 10. These sets often have a built-in catalog of interesting astronomical objects and pointers which direct the observer which way to slew the telescope to reach the object. Other, more-sophisticated systems allow a computer to be interfaced with the telescope for completely automated control.

7

Observing with friends

The first friends I observed with were, of course, my family.[151] On camping trips as a child I would view the night sky, not really comprehending what I saw but finding it intriguing. After a little schooling and a few years I found myself as a professional astronomer in a well-equipped observatory. But I observed with colleagues, not necessarily friends.[152] Our attentions were often diverted by competition for budget funds and office politics. At a later stage in my career I found myself designing optical instruments and data systems but not observing at all. And so, I committed the unpardonable crime (in some professional astronomer's eyes) of purchasing a small telescope (a 20-cm S–C in fact) and becoming an 'amateur' astronomer. I even joined an astronomy club and therein found kindred souls.

The astronomy club environment

Astronomy clubs have suffered an image problem in recent years. They often attract rather single-minded loners with an overbearing passion for one thing – astronomy. The nature of the telescope as an instrument encourages single-observer projects, for there is usually only one eyepiece per main telescope. Before the advent of government sponsored team-research astronomi-

---

[151] My own children have shown considerable interest in astronomy but I have not projected my own wishes upon them. I am still not sure if they have a deep and abiding interest in the stars or if it's the fact that when we drag the telescope out they get to stay up late and have Daddy's undivided attention.

[152] The vast majority of professional astronomers I have observed with are life-long friends. As in any profession, however, there are one or two whom I find most useful as examples to warn my children about. I do find the incidence of truly vile characters in the astronomical community to be less than in the general population – but perhaps I idealize my own profession.

cal projects, astronomy was largely conducted by rich or self-sacrificing individuals not well understood by their contemporaries.[153] Indeed some rather odd folks have set the foundations of astronomy in place.[154] Thus, astronomy clubs, like computer and chess clubs, have often been called refuges for nerds.[155]

Astronomy clubs are full of sensitive, caring people – it's just that their caring extends to the observable limits of the Universe. And this may stretch their sensitivity a bit thin back here on Earth. I find the denizens of astronomy clubs to be a hearty, robust group who may have only an interest in astronomy to bind them together but that is enough. While there are some individuals in my clubs (I belong to three)[156] who may hold despicable political or social views, we don't discuss them out of mutual respect – but I find their observing tips helpful. And I want to hear of their observations and thoughts upon the cosmos.

What happens if there is not an astronomy club nearby? I hereby nominate you as the charter president of the <insert geographical name here> Astronomical Society! The nominations are closed. All in favor of adopting the ticket by acclamation, say, 'Aye'.

'Aye!'

The Ayes have it.

That was simple, wasn't it? All you have to do now is find some other astronomers. Ask your local high-school science teacher or librarian. They always seem to know who's doing a little stargazing.

Now that you're in an astronomy club, what do you do? Each club sets its own agenda but the two obvious choices are to have

[153] Consider society's view of the classical astronomer. Who else works all night and sleeps all day except burglars and ladies of questionable merit?

[154] As a child I was told that there was a crazy man living nearby who listened to the stars on his radio. Only later did I discover that he was Grote Reber, now considered the father of modern radio astronomy, who built the first dish-type radio telescope with his own funds and got radio astronomy rolling. For a description of his instrument, see *Unusual Telescopes*, Peter L. Manly, Cambridge University Press, 1991, p. 77. How's that for a blatant advertisement for one of my other books?

[155] I find the term "nerd" demeaning, for it connotes a narrowly focused, socially challenged individual wearing thick glasses and a plastic pocket protector. I am not a nerd and have never worn a pocket protector. I am a full-fledged card-carrying Techno-Geek!

[156] East Valley Astronomical Club, Phoenix Astronomical Society and the Saguaro Astronomy Club. I first joined the San Jose Astronomical Association but I don't live there any more.

Figure 7.1. Astronomers setting up for a club star party.

meetings and observe. Observations are the fun part, in my opinion. I enjoy wandering around the telescope field in the dark, chatting with other astronomers and seeing what strange objects they've found in the sky. Quite often interaction at a star party really amounts to one-on-one instruction in the skills of observing and the lore of astronomy. I feel I never would have learned as much as I have if I'd had to do it alone. The social interaction occurs even more after the observing session when astronomers gather at a coffee shop to warm up and exchange information.[157]

Meetings usually involve a presentation by a member or visiting astronomer. The business of the club will also be decided here.[158] As in any volunteer organization, this is where politics can erupt and spoil the fun of belonging to a club. Remember that many astronomers are loners and have not developed a high

[157] On the rare occasions when clouds gather on the afternoon of an Arizona observing session, some club members will telephone others and arrange to meet at a restaurant without even going to the observing site first. We don't even pack our telescopes in the car on such nights – but we do get quite a bit of astronomy done.

[158] In order to assure that meetings were primarily concerned with astronomy and not small club politics, one club made it a rule that no more than 20 minutes of each meeting would be taken up with business. Any squabbling about policy from the floor generally resulted in the disgruntled parties being appointed as a committee to solve the problem and report back at the next meeting.

level of skills in conflict management.[159] One club officer quipped that he felt like the president of 'The Association of People Who Don't Usually Join Clubs And Shouldn't'. On the other hand, for each obstreperous astronomer there usually appears a peace-maker who can iron out the wrinkles in club life.

One of the great strengths of belonging to a club is the possibility of participating in joint projects. On the East Coast of North America and in Europe this usually takes the form of a large, club-owned telescope. Several meter-class instruments have been built and at least one telescope of 1.5-meters aperture is currently under construction. Many club telescopes are as fully capable as professional grade instruments.

In the central and Southwestern areas of North America, club projects tend more toward organizing special observations or meetings of astronomers. The yearly Texas Star Party and Riverside Telescope Maker's Conference are examples. Our local Arizona clubs also have a yearly pilgrimage to Kitt Peak or one of the other world-class observatories dotting the mountains in our state.

Field observations

Before I could use my telescope in the field, I had to purchase several accessories. The first, as mentioned in Chapter 6, is a drive corrector. Then, in order to power the drive corrector, I bought a battery.[160] Although many observers use their car batteries without problems, more than one has observed all night with a weak battery and then found that he doesn't have enough battery energy to start the car in the morning.[161] Batteries can be dangerous when tumbled over in the dark so I bought a sealed battery box. Add in power cables, connectors, a battery tester

---

[159] Insight into the motivations of amateur astronomers and other professions can be found in *Amateurs: On the Margin Between Work and Leisure*, Robert A. Stebbins, Sage Publications, 1979.

[160] I carry a separate 12-volt battery specially designed to be slowly drained of all its energy at a rate of 1–5 amperes, called a "deep draft" battery. It was originally developed to power trolling motors on fishing boats. The usual car battery is designed to deliver 100–300 amperes for 5 seconds and discharging it completely damages the battery.

[161] This is one of the reasons that I try to avoid observing alone and if I have to, I let people know where I'll be and when I plan to return. There is also the fact that the mountains and deserts of Arizona can be lethal to the unprepared. In addition, we have critters in the woods who might regard astronomers as either invaders or dinner.

and battery charger. Some times these little additions never seem to end.

The accessory box is something you'll evolve as your needs change. It will usually seem too small for all of the eyepieces, wrenches, spare batteries and other junk which you will collect.[162] I keep my accessories in a metal case originally designed to hold ammunition. It is dust and water proof with a sealed lid. Such cases are available at military surplus stores. I have lined the box with foam rubber to protect the eyepieces when my van jolts along primitive roads. With all of the paraphernalia associated with a field expedition, I have at times forgotten to pack some critical piece of equipment.[163] In order to prevent future incidents, I made a telescope packing checklist to ensure that I loaded all the required bits and pieces in the car. This also assures that items such as an extension cord or flashlight which might be borrowed between observing sessions from the pile of telescope equipment are returned before you depart. The checklist is shown in Appendix 6.

I enjoy taking the telescope out to a dark site occasionally. The trip is usually combined with a camping expedition and often it includes the company of the local astronomy club, affording pleasant companionship. The camping allows me to work all day, observe remotely all night and not have to drive home while sleepy. I know of some observers who even pack a portable darkroom and develop their pictures at the observing site.

Observing portably can also be done closer to home in order to bring the telescope to the viewers. Many astronomy clubs present public observing sessions, usually in municipal parks. While the lights of the city may hinder viewing some of the more obscure nebulae, Lunar and planetary views are usually enough to pique the interest of most people. I have encountered two potential problems while demonstrating my telescope in public sessions. First, the cables and electrical cords will usually be snagged by unsuspecting people who are not used to operating

---

[162] I was once foiled in my observations by lack of a special tiny wrench needed to tighten a camera fitting. The next day I purchased one and put it in my accessory box. In the subsequent fifteen years, the fitting has remained tight and the wrench has never been used. But I know that if I remove the wrench from the box, the fitting will automatically loosen. Keeping the unused wrench is a form of preventive maintenance.

[163] All right, I'll admit it. I forgot the tripod. As it happened, we were clouded out that night but when I returned home and found it in the garage I felt rather foolish.

in the dark. In order to prevent this I place the power battery directly under the telescope tripod and secure loose cables with gaffer's tape. The second problem occurs when a line of children is waiting to view. Invariably there is horseplay among those in line. I've always feared somebody shoving the observer, causing him to poke his eye with the eyepiece. Thus, I place myself between the observer and the next person in line. My proximity also seems to have a calming effect on the first few children in line.

The problems of observing with children are more than out-weighed by their enthusiasm and interest. Our local astronomy club has staged a series of special observing sessions for many schools, usually involving the school science club or special edu-cation class. Children are always full of questions and most of them, such as how far the Moon is, can be fielded easily. There are, however, in each class, one or two little geniuses who can manage to come up with at least one of the Fundamental Questions still left unanswered in astronomy. Amazingly, it appears to take them only about ten minutes of observing to for-mulate their questions.

Regional, national and international networking

You may feel a need for a bit more contact with astronomers than you can find within the local astronomy club. Often clubs belong to umbrella organizations which coordinate inter-club activities and ensure that two nearby clubs don't schedule major activities on the same night. There are also national organizations such as the Astronomical League or the British Astronomical Association. A list of such organizations is given in Appendix 1. Some national and international organizations are built around a particular type of observation such as variable stars, eclipse chas-ing or occultations. Their names are also given in Appendix 1.

If you still hunger for astronomical contact, there are two options left. One is the computer bulletin board and the other is the nearly lost art of letter writing. In most astronomy magazines you will find names and addresses of people wishing to corre-spond. Often they are in some remote section of the world and cannot find another astronomer within several days' travelling distance. Years ago I corresponded with a young astronomer overseas and enjoyed his letters for several years. Several astronomers in my local Arizona club corresponded with an Australian and then finally decided to visit him, travelling halfway around the world just on the strength of a few letters.

8

Projects

In previous chapters serious projects such as asteroid occultations were discussed but these tend to be one-shot events that occur months or years apart. The same is true for eclipse chasing. There are, however, several programs which involve regular observations on a weekly or monthly basis. These are long-term projects in which the astronomer invests time and dedication, often working for years before finishing the effort. Indeed, some of the programs have no discernable end and are in the nature of patrols for events like supernovae or gathering of data for professionals to use. The ancients may have believed the Heavens to be constant and unchanging but we find a dynamic Universe in which objects are constantly in motion with respect to esch other, stars are born and die, comets appear as if out of nowhere and meteors blaze for a moment of glory.

It has often been said that astronomy is one of the few sciences where amateurs can make a professional contribution. The professional astronomy community has always had more observing problems and astronomers than telescopes to go around so they welcome amateurs.[164] Of course, amateurs must choose observing projects within the capabilities of their admittedly smaller instruments but it turns out that there are many small-telescope projects looking for a home.[165]

---

[164] Most professional astronomers typically have only two or three observing sessions per year for a week at a time on a major telescope. Many are envious of amateurs who have unlimited access to their own private instruments. In 1986 I observed on 111 nights of the year, a total which few, if any, professionals could claim.

[165] Three books that explain many potential projects for amateurs are: *Observational Astronomy for Amateurs*, J. B. Sidgwick, Enslow Publishers, 1982, ISBN 0 89490 067 6; *Experimental Astronomy for Amateurs*, Richard Knox, St. Martin's Press, 1976; and *Projects and Demonstrations in Astronomy*, Donald Tattersfield, Wiley & Sons, 1979, ISBN 0470 26715 1.

Planetary, Lunar and Solar patrols

Under the sections on each of the nearer planets, mention was made of sketching or photographing the weather patterns. While it is an interesting personal study to chart the motions and changes of features, some observers have banded together to share information and disseminate tips on how best to record planets and the Sun. The Association of Lunar and Planetary Observers (ALPO) and the British Astronomical Association (BAA) have sections and publications devoted to these studies. Their results compliment professional observatories which may be clouded out during important changes. Since most astronomers belonging to these organizations rely on visual observations, a case can be made where greater detail (at the expense of possible subjectivity) can be seen than that using photographic techniques. This is, at times, a controversial subject since several visual observers have claimed to see features which were later shown to be absent when spacecraft visited the planets for close-up images. On the other hand, as in the case with Saturn's hundreds of separate rings, the visual observers proved correct.

At times these organizations have studied issues which are controversial or even scoffed at by some astronomers. Typical among these is the ashen light of Venus, as mentioned in Chapter 3. There are few areas of study more controversial, however, than Transient Lunar Phenomena (TLP). The question revolves around whether the Moon is still actively forming or is a dead rock. Of course, the surface of the Moon will be changed by meteor impacts but that is not the main issue.[166] Several observers have reported bright areas, often in craters, which they have interpreted as either outgassing or volcanic activity. Others have reported color changes and regions which appear hazy while surrounding areas remain in sharp focus. Finally, bright, flashes have been observed, especially on the dark side of the Moon where pale Earthshine illuminates the surface.[167]

---

[166] A vivid description of an observed meteor impact in 1187 is given in *Cosmos*, Carl Sagan, Random House, 1980, p. 85. A comparison of photographs over the last century, however, does not reveal any new impact craters.

[167] For a more complete list of Transient Lunar Phenomena see *The Sky: a User's Guide*, David H. Levy, Cambridge University Press, 1991, p. 83.

While I would encourage observers to partake in the studies, realize that it is a slow and tedious process. I have stared at the Moon for hours, made copious notes and seen nothing really out of the ordinary. But I have also listened raptly to excited observers who have seen something. The frustrating part of their tales is that nobody else happened to be looking at the Moon at that time and reported a corroborating observation. Thus, it remains an unsubstantiated visual observation, suspect because of the well-known susceptibility of the human eye to various optical illusions. One cure for this is to maintain an observing program within the guidelines of one of the above organizations. That way, if something is seen and it is real, then one of the other scheduled observers may also report it. The best time to look is probably when the Moon is at old or new crescent phase and the dark side is more brightly lit from Earthshine.

Variable stars

In Chapter 5 one type of variable star, the eclipsing binary, was discussed. There are other methods by which stars change their brightness and sometimes even their color. Many giant and supergiant stars vary in brightness in a regular and repeating way. These are called periodic variables. A complete cycle of brightening and fading may be as short as a few days or as long as several years. Other stars may suddenly brighten at irregular intervals for a few days. Then they will remain faint for months or years before unexpectedly flaring again. Variable stars of all types have been identified and catalogued by the American Association of Variable Star Observers (AAVSO) and the Variable Star Section of the British Astronomical Association (BAA). Thousands of them are well within the grasp of a 20-cm S–C telescope.

Measurements of the brightness of variable stars can be made either visually or photoelectrically, as described in Chapter 10. Visual observers must learn to recognize small differences in brightness by comparing two or more stars. A typical observation of variables starts, as all serious observations do, with the astronomer noting in his log the weather, date[168], time of observa-

---

[168] Variable-star observers use the Julian Date, the number of days since the arbitrary starting point of January 1, 4713 BC, at Noon, Greenwich Mean Time. This avoids the complexity of changing calendar systems when calculating intervals between events.

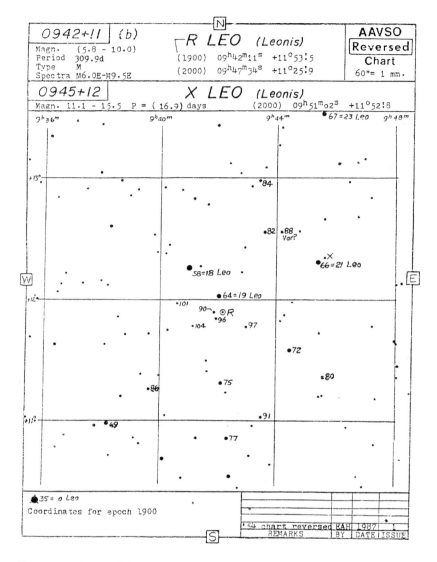

Figure 8.1. Typical finder chart for a variable star. Illustration by special permission from Janet Mattei, American Association of Variable Star Observers.

tion and the configuration of the telescope. The observer will have a set of finder charts for the target stars and specific instructions from the AAVSO, BAA or some similar organization, as shown in Fig. 8.1.[169] After examining a couple of familiar stars to assure that the seeing is adequate, the observer is ready to make a visual estimation of stellar magnitude.

The finder chart for the variable star will also contain two or more comparison stars of known visual magnitude, similar in brightness and color to the target star. There are two methods for estimating star brightness; the step method and the fractional method. In the step method, the observer must first have learned, by training his eye on practice stars, to recognize a brightness difference or step of 0.1 magnitude (some observers use 0.2 magnitude). The brightness of the target star is then estimated to be so many steps brighter or fainter than the comparison star. In the fractional method two comparison stars are used, one brighter than the target star and one fainter. The observer estmates the fraction of the brightness interval between the comparison stars that the target star shines. Of course, this method requires more comparison stars and considerably more observing time but it is preferred by most people.

When estimating brightness the observer should always use the same place on the retina, for in some eyes the relative sensitivity varies from point to point on the retina. Many variable stars are red and for red light there is an apparent brightening of the star if it is stared at for a few seconds. This is similar to a time exposure in photographs and is called the Purkinje Effect. While this may allow you to see faint stars a little better, it destroys the ability to compare brightnesses. Observe the target and comparison stars with quick glances, lasting just a second or so. It will take many glances at all three to settle on a magnitude estimate. When observing a variable as a part of a regular program do not look up the star's brightness from the last measurement before observing it, lest the observation be biased by the expectation of seeing it at some particular magnitude.

Of course, always record your observations clearly in the observing log. After each session, review your notes to see that

---

[169] The AAVSO produces special finder charts for observers who use S–C telescopes with a 90° mirror or "star diagonal" which flips the image left to right. The maps are printed reversed so that the chart matches the eyepiece view.

they are coherent and complete. Completeness includes notes on the weather, seeing and transparency of the sky, possible interference by Moonlight and anything else that could affect the observation such as a new eyepiece or realignment of the mount. I once discovered some anomalous data points months after the observation and puzzled over them until I reread the observing log for that night and saw an entry that I'd stopped observing early because I didn't feel well. In fact, I was coming down with a first class head cold which, for some reason, usually impairs my night vision. Thus, I was able to understand the anomalous data and remove them from the reduced readings.

Visual variable-star observers are often paired with other observers in distant lands. The first reason for this is that the organization coordinating the variable-star program compares the two as a check on the quality of the data. The reason for pairing distant observers is that if one is clouded out with some weather pattern, the other is unlikely to experience the same pattern and thus will make up for gaps in his partner's data. Pleasant and long-lasting correspondences have been built out of such relationships. Many amateurs have also formed friendships with the professionals who use their data and often they are listed as co-authors on the scientific papers that result.

Lunar occultations

I have been involved for several years with the International Occultation Timing Association (IOTA) making observations of the Moon. There are basically two different measurements made; the graze and the total occultation. In a graze the Moon sweeps past a star and either the North polar regions or the South polar regions just barely cover the star. For a few moments the star will be hidden behind some Lunar hill and then it will shine through a valley before disappearing again. The process can result in a score or more of events as the star winks and blinks in the eyepiece. By recording the time and duration of such events, a profile of the Lunar topography can be built up. Since the Moon appears to rock from side to side and tip the Polar regions toward and away from us by a few degrees (Lunar libration) as it moves around its orbit, then quite a few areas can be mapped by observing grazes.

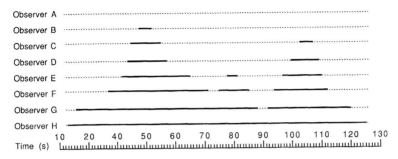

Figure 8.2. Typical graze results. Dark lines indicate times when star was not visible. Vertical scale is exaggerated by a factor of 500.

Typically, several observers are stationed a few tens of meters apart along a line pointing toward the azimuth of the Moon at mid-graze. Each observer then sees a different 'slice' of the Moon. The times of disappearances and reappearances are measured using WWV radio time signals. Usually, the observer speaks into an audio tape recorder with the radio time signal playing in the background.

When the graze happens, some observers can be surprised with the speed and number of events and thus become flustered. In order to prevent this, few words should be spoken on the tape and only then to record specific events. The words 'out' and 'in' are used to describe the presence of the star's image. 'Fade' refers to the star dimming without completely disappearing and 'wink' means that the star reappeared and disappeared with a duration too short to be discerned. After the observation, the times of particular events can then be determined from the recording. The reaction time of the observer must be subtracted from the recorded time in order to obtain accurate data. This generally runs about two tenths of a second for most observers but it can be seriously lengthened by fatigue or faintness of the star.

The graze is obviously a team sport but during the observation each astronomer sits alone, perhaps within shouting distance of the next observing station. As the moment for the event approaches and the star nears the limb of the Moon, there is a bit of excitement in the air. If one listens closely, sometimes the shouts of adjacent astronomers watching lower-altitude slices can be heard. On one such graze I could hear the call of 'Out!' from an observer three stations away. Then a friend at the next nearest station called 'Out!'. I pictured in my mind the shadow of some Lunar mountain rushing across the nearby fields as the next station called 'Out!'. In my eyepiece the star held steady until I heard the next station call 'In!' and then I realized that the shadow of the very tip of that mountain fell somewhere between our two stations (see observer A in Fig. 8.2). While it may seem as if I took no data, in essence I had measured an upper limit on the altitude of that particular hill.

Total Lunar occultations occur when the limb of the Moon completely covers a star. This is most easily done at the dark limb since the Sunlit limb is very bright and glare often masks faint stars. Between new Moon and first quarter observations are made in the early evening of stars which disappear behind the Eastern limb. Between third quarter and new moon observations are made after midnight of stars which reappear from behind the Moon at the Western limb.

Occultation observations are made for several reasons. First, noting the limb position with respect to an accurately measured star allows a calculation of the position of the center of the Moon at the instant of the observation. This aids in understanding the motions of the Moon in its orbit, which is not a simple matter. An accurate theory of Lunar motion including gravitational effects from the Sun and all the other planets plus relativistic effects is still being pursued. A second use of the data is to determine the precise shape of the Moon itself including surface features at the limbs. A third use for the data occurs if the star happens to be a double star, in which case it may reveal itself when it fades partially and then a second or two later disappears altogether. Many stars previously unsuspected of being doubles have been discovered this way, as discussed in Chapter 10 in the section on photometry.

For a typical Lunar occultation the observer works from a table of predicted events with a set of stars whose positions are known precisely. The table is calculated for the observer's postion on the Earth and the times are accurate to within a few seconds.[170] The object of the observation is to accurately time the event and thus refine the calculation of when the event occurs. For a dark limb

---

[170] Tables are available from the International Occultation Timing Association (IOTA). Their address is given in Appendix 1.

disappearance this is not difficult. As with a graze, the observer notes the event with an audio tape recorder and WWV radio time signals.

For a reappearance the situation is a bit more difficult. The observer has nothing to concentrate on until the star appears. The predictions list the Lunar latitude at which the star will reappear but unless you have a protractor reticle eyepiece (and I don't) you will have to estimate the position on the limb. Occasionally the Earthshine will be bright enough to allow the observer to see specific Lunar features and get his bearings but more often than not, the dark-limb is just a gray smudge.

I generally use an eyepiece with high magnification so that the background sky brightness is less and thus I can see fainter stars. Even if the star is initially bright, it may have a faint companion which will be revealed when the bright one disappears. The background sky brightness is also aggravated by the fact that less than half a degree away is the Sunlit portion of the Moon. Any water vapor or dust in the air will cause the sky for a degree or more around the Moon to glow with scattered light.[171] I do not use my highest-power eyepiece when observing occultations because it has poor eye relief and my eye fatigues easily when using it for prolonged periods. Fatigue, as mentioned earlier, will wreak havoc on the data when trying to estimate your reaction time. A magnification of about 150×–200× gets the Sunlit portion of the Moon out of the field of view and still allows comfortable viewing.

Mutual Jovian eclipses

In Chapter 3 the Galilean satellites of Jupiter were mentioned. They whirl around the giant planet like a miniature Solar system. Although they all orbit in approximately the same plane, each moon has its own unique orbital plane. Thus, it is possible for one moon to eclipse another as seen from Earth. It is also possible for one moon to cast a shadow on another. The moons regularly disappear into Jupiter's shadow and reappear from it. If the

---

[171] This light is often brighter than reflected city lights. One of the more prolific IOTA observatories is located in the center of a city of more than a million people. The observatory director chose to pursue a program of Lunar occultations because the city lights wouldn't make much difference when compared to the Moonglow.

timing of these events were recorded, then more precise orbits for the Jovian moons could be calculated. The Association of Lunar and Planetary Observers (ALPO) runs a special observing program to make these measurements.

The 20-cm S–C is an ideal telescope for these observations since it can easily show all of the Galilean satellites. Typically, mutual events are predicted well ahead of time and observers are asked to determine the moment at which an event happens. Moons moving into or out of Jupiter's shadow will take several minutes to cross from the darkest part of the shadow to full Sunlight. It takes a keen eye to discern the instant when a satellite first begins to fade or stops increasing in brightness. Thus, an estimate is made of the time at which the satellite is half as bright as it is in full Sunlight. This can be aided by comparing the satellite in question with the brightness of other Galilean satellites or nearby field stars. As with other occultation measurements, the time is recorded from WWV radio time signals.

9

Photography

Whole books have been written on astrophotography.[172] I don't mean to supersede them here but merely to interest the new astronomer in a fascinating aspect of observing. While not much equipment is required to get started, be warned that making star pictures can consume vast amounts of time and effort. But then again, that's what most people are looking for in a hobby. Within this book I have listed, for each astrophoto, the exposure time, optical configuration and film type so that the reader may get a feel for typical photographic parameters. While some professionals state that photography is a science, determining the correct exposure is usually a matter of trial and error. Framing the subject and composing the ideal picture is certainly an art.

Since a 20-cm S–C telescope is already at hand, all that is necessary is a camera. The common 'idiot proof' auto-focus, auto-exposure, point-and-shoot cameras usually won't work for astrophotography. Their light meters aren't designed for the low light levels involved and if you bolt one on the backplate of your S–C the range sensor will see only the rear of the telescope and tell you that you're standing too close to your subject. Most photographers use a 35-mm camera body. Mine is a 20+ year old junker whose light meter hasn't worked in a decade.[173] The shutter doesn't work except on the 'Bulb' or 'Time' (for time exposures) setting. At least it has a working winder and a flange on the front that I can use to attach the camera to the telescope. No other sophistication is required in a beginner's astrocamera.

---

[172] See *Astrophotography for the Amateur*, Michael Covington, Cambridge University Press, 1985.

[173] Some of the more automated 35-mm cameras require battery power to open and close the shutter even if the camera is operated in a 'manual' mode. Using exposures of an hour or more may drain the batteries completely, causing the shutter to either prematurely close or, worse yet, fail to close at the end of the exposure. Test your camera in the cold night air before trying it on your telescope.

Before removing the lens from your 35-mm camera and attaching it to the telescope, try a simpler type of astrophotography. Secure your camera with its wide-field-of-view lens to the accessory bracket (piggy-back mount) of your telescope. Open the f stop on the lens as wide as possible and set the focus to infinity. Set the drive on your 20-cm S–C going and point the telescope/camera to some interesting area in the sky. Anywhere in the Milky Way will generally yield good results. This is called piggy-back photography and the telescope itself isn't used except as a rotating platform which tracks the stars. Try exposures of 1, 2, 5 and 10 minutes.

You should wind up with some photos which show many more stars than you can see with the naked eye. If you live near a city, you may see a green or gray foggy background in the longer exposures. You should also see much more vivid colors in many stars than you see with the naked eye. The form of the Milky Way may also be more clear and perhaps the Great Rift or the Coal Sack might be discernible as a darker cloud in front of the myriad stars of our local galaxy.

A few quick experiments with piggy-back photography will also yield a better understanding of how various films react to star images. This is especially important if you are developing your own films and perhaps even making prints. Since it is much easier to make a series of unguided wide-field test shots than hours-long exposures just to experiment with various films, many beginning astrophotographers spend some time with piggy-back photography before ever attaching the camera to the back of the telescope.

You may discover that some short-focal-length lenses with low f numbers yield crisp images at the center of the field of view but at the corners the stars become fuzzy or have elongated shapes. This is the mark of a cheap lens.[174] The defect can be minimized by setting the f number just a stop or two higher. You will have to compensate for the smaller aperture by exposing the film for a longer period but extending a two-minute unguided wide field exposure to eight minutes isn't a critical expenditure of time and the resulting improvement in image clarity may well be worth it.

---

[174] More correctly, this is the mark of a cheaply made lens. I own one particularly bad example which was not cheap in terms of purchasing cost.

At this point you will discover that astrophotography involves one trade-off after another. In lens or telescope selection, you can either have a wide field of view at low angular resolution or a narrow field of view at higher resolution. You can't obtain wide fields of view at high resolution without spending astronomical sums of money.[175]

A second trade-off is in film types. The primary experiment among films is to look at differences in speed (sensitivity) and resolution. More-sensitive films with higher speed ratings have an advantage in that they require a shorter exposure in order to capture the same number of stars as a slower film. Unfortunately, many high speed films have a grain or resolution limit governed by the size of the photosensitive particles in the emulsion. For some films this limit may be much worse than your camera/telescope combination or even the atmospheric seeing. A secondary trade-off is that higher-speed films and higher-resolution films are usually more expensive.

Another experiment is an examination of the reciprocity failure phenomenon. While one might think that doubling the exposure time should allow images half as faint to be captured, this is not so for exposures longer than a few minutes. Without going into the chemistry of the photographic process, suffice it to say that the sensitivity of an exposure is nonlinear with time. Doubling the exposure time may yield only a factor of 1.5 increase in sensitivity. The next doubling of exposure time might add another factor of 1.2 or so. Soon a point of diminishing returns is reached where the astronomer's patience is more of a factor than gaining that last half percent of sensitivity. Photographers have developed several techniques to forestall the reciprocity failure effect such as chilling the emulsion to the temperature of dry ice or pre-soaking the film in exotic mixtures of gases.

The phenomenon of reciprocity failure becomes particularly interesting in color films where the blue, green and red emulsions may fail at three different rates. I have some curious photos of the Orion Nebula taken several years ago showing the entire

---

[175] You will also run into the law of diminishing returns which states that given an inexpensive optical system it may cost twice as much to obtain twice the resolution but it will cost ten times as much to double the resolution again and a hundred times as much to double it a third time.

object as a brilliant red. The nebula isn't really that color but the red emulsion was the last to fail in that particular film. In recent years, emulsions have been improved and the problem appears less often.

The question of color film versus black-and-white film is another trade-off. In general, color films have less sensitivity and resolution than black-and-white emulsions. While color exposures have fewer scientific applications in astronomy, they can produce pleasing and, at times, spectacular views of faint objects invisible to the naked eye at the eyepiece. In addition, they can show the individual star colors. Special high-resolution films with extended sensitivity in either the ultraviolet or near infrared have also been developed especially for astronomers. Such specialty films, while a bit more expensive, can be ordered from most camera stores and they usually produce much better astrophotos than consumer-grade films. It doesn't really matter which film you try first as long as you record your exposure times, film types, telescope set-ups and subjects and learn from your results.[176]

Once you understand the basics of film, exposures and planning a photography session, it's time to bolt the camera on the back of the telescope and use the 20-cm aperture optics. Many astrophotographers support the camera on a separate tripod and adjust it until the camera is looking into the optical train. Of course, as the telescope slowly tracks the sky, the secondary tripod must constantly be adjusted. I believe the two-tripod system was originally developed by and for people with Newtonian telescopes on spindly German-type mounts which couldn't support the weight of a camera at the end of a long moment arm near the eyepiece. Every commercially made 20-cm S–C I have seen is designed with provision to mount a 35-mm camera body directly to the telescope backplate.

Many different mechanical and optical arrangements are possible between the telescope and the camera body. Initially, a prime-focus coupling would be easiest and simplest, as shown in Fig. 9.1. The required adapters can be purchased at camera stores

---

[176] While it may seem silly to record the name of the object you photographed since the object should be recognizable, when you get some blank slides or negatives back from the developer, the purpose of this record should become clear.

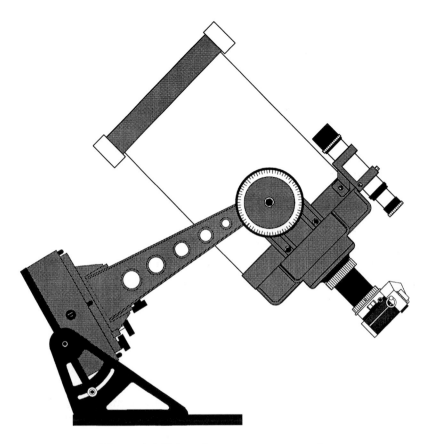

Figure 9.1. Prime-focus camera mounting.

or from the manufacturer of the telescope.[177] The camera will throw the balance of your telescope off. Thus, you must add counterweights to first balance the declination axle (in two directions) and then the polar axle.

A 20-cm aperture f/10 telescope obviously has a 200-cm focal length. This implies a plate scale of about 102 arc seconds per millimeter. The standard 35-mm × 24-mm negative size of a 35-mm camera thus yields a field of view for the system of 0.98° × 0.68° with a 1.2° diagonal. Before delving into exposures of an hour or more of faint nebulae, may I suggest an exercise which will

---

[177] Beware, however, because there are many similar-appearing camera body adapters. In an attempt to adapt cameras, focal reducers/extenders, filters and other accessories, you will soon wind up with a frightening collection of adapter rings.

quickly familiarize you with photography through the telescope? Take a picture of the Moon. You won't have to guide the telescope for long periods and the process will allow you to find the best methods for focusing and aligning your particular camera.

After attaching your 35-mm camera body to the prime-focus telescope adapter, you will obviously point the telescope at the Moon and focus it. Most 35-mm camera viewfinders have a ground glass focusing screen which facilitates adjusting the optics for a sharp image. Some have a central area with a split-image feature allowing fairly precise focusing. When using fine grain focusing screens on nights of very steady seeing, however, it is difficult for some observers to judge where exact focus occurs. Most camera manufacturers make an accessory magnifier which fits on the viewfinder and doubles or triples the apparent magnification of the central portion of the field of view. The magnifier flips out of the way when not in use, as shown in Fig. 9.3. I have found this accessory to be very handy.[178]

The Moon is bright enough for most camera exposure meters to register an appropriate shutter speed. Beware, however, that some cameras measure light over the whole field of view and some cameras measure only in the central area. With the Moon centered, about half the area of the field of view is black night sky and the rest is bright Moonlight. Most exposure meters were designed to deal with normal, lower-contrast terrestrial scenes and not the very-bright/very-dark scene now in the viewfinder. Therefore, take the indicated or recommended shutter speed as a suggestion only. Use that exposure but also take shots at twice, four times and eight times the indicated exposure. Similarly, try 1/2, 1/4 and 1/8 the suggested shutter speed. This is called bracketing the exposure. It consumes a bit more film but later, when dealing in exposures that are minutes or hours long, such experimentation and its results will come in handy.

When taking a few of the exposures, look through the viewfinder and, just after the exposure, examine the image for

---

[178] For those viewfinders with a split-image center section, focusing on a star may be difficult as the split-image optics are made for focusing on a vertical line. A point-source image such as a star can be focused using the split-image feature by setting the camera orientation so that the top or bottom of the frame is aligned with North. Then nod the telescope up and down in declination slightly, causing the star to move between the upper and lower halves of the split image. If the star shifts from side to side as it crosses the center line then the telescope isn't focused properly.

vibrations. In some older cameras such as my own, the mechanical motion of the camera's viewfinder mirror and shutter produce vibrations in the telescope which will blur the image during the exposure. My camera has a control which allows me to manually move the focusing mirror up out of the light path and lock it in place. I can wait a few seconds for the vibrations to dampen and then take my picture but the shutter itself still produces enough vibration to blur the image slightly. You will have to experiment with your own camera to see if this is a problem. If it is, then you can use the 'black card' trick. Set the camera to a 'bulb' or 'time exposure' setting. Hold a black cardboard sheet which is larger than the aperture of the telescope in the light path in front of the telescope.[179] Be careful that the card does not touch the telescope and thus introduce more vibrations. Using a cable release, open the camera shutter and wait several seconds for the vibrations to dampen down. Then swiftly remove the card from the light path for the required exposure time. Granted, it is difficult to accurately calibrate your wrist to exactly a 1/4-second exposure but most astrophoto shots will be well over one second.

Having experienced at least one astrophotography session with the Moon, you will probably be disappointed at initial results. Fear not, for you are in good company. Only one frame of 36 contained a discernible image in my first roll and that was out of focus and blurred. I never even showed anybody my results until about roll no. 15 and the earliest astrophoto I ever had published is from roll no. 107. In spite of this, I enjoyed the learning experience and reveled in the idea that I was pushing back the frontiers of Mankind's knowledge of the skies. Well, perhaps I wasn't exactly on the forefront of experimental astrophotography at the time but it seemed so and it was fun.[180]

For your second astrophotography session I'd recommend the Moon again. After reviewing your results from the first session,

---

[179] See *Astrophotography for the Amateur*, Michael Covington, Cambridge University Press, 1985. Author Covington states that the black back cover of the book was designed to be used as a 'black card' for astrophotography.

[180] One possible diversion can occur at this point if you become involved deeply in developing your own astrophotos. Setting up a darkroom, fooling around with chemicals and making enlargement prints can consume all of your astronomy time. It happened to me and only with a great exercise of willpower was I able to wrench myself away from the fascinating technology of the darkroom, buy a CCD TV camera and become equally trapped by the fascination of computer image processing software.

look at the photos and your notes. If you don't understand why each photo was underexposed or overexposed then at least you'll learn to take better notes during the next observing session. For the second session you might want to try either a tele-compressor or a tele-extender or tele-converter; as shown in Fig. 9.3. These are small lenses which mount between the telescope and your camera. Depending on the design, most of them either double or halve the effective focal length of the telescope. A focal reducer (also called a tele-compressor) thus allows your telescope to operate as if it were a 20-cm f/5 optical system. The field of view will be twice as wide as before and exposures will be approximately one quarter as long as before since exposure time is proportional to the square of the f number. While the shorter exposure times may be more convenient, the Moon's images will be only half the size as before.

Similarly, a focal extender lens (also called a tele-extender) makes your telescope into a 20-cm aperture f/20 telescope. The Moon's image will probably overfill the field of view. Exposure times, however, will be four times as long. This is a typical astrophotography trade-off where you will be able to record greater detail on the Moon but only by exposing the film for longer intervals.

Once you have mastered the art of photographing the Moon (or have grown sufficiently tired of the exercise), one more experiment is possible. By this time, I assume that you have run five to ten rolls of film through your camera. You have learned not to bump the telescope during a long exposure. You have learned to wind the camera before each exposure. You have learned to set the shutter speed before each exposure. You have learned to record the data for each exposure. You have learned to bring all of the needed accessories to the telescope at the start of each observing session. You have learned to plan the session with clearly stated and well thought-out written objectives. I assume this because you can't have gone through five rolls without making some of those errors. I know I had to learn those things the hard way.

Just as the tele-extender and the tele-compressor can vary the effective focal length of the telescope, your collection of eyepieces can also be used to add magnifying power to the telescope, as shown in Fig. 9.2. An eyepiece projection kit composed of an eyepiece holder and a set of hollow tubes is required. The

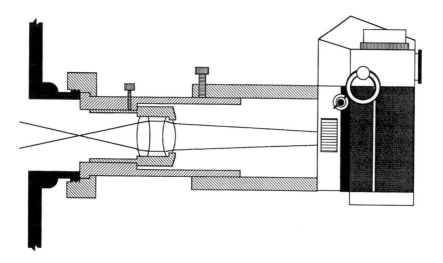

Figure 9.2. Eyepiece projection photography.

eyepiece slips into the holder and the camera attaches to the last of the hollow tubes.[181]

While the eyepiece usually converts the converging rays from the telescope into parallel bundles of light which your eye will interpret as an image located at an infinite distance, when used as a projection lens the eyepiece will re-focus the scene at the camera's image plane at a larger plate scale (magnification) than if the camera were placed at prime focus.[182]

The exact magnification depends on the focal length of the eyepiece and the distance between the eyepiece and the camera. Thus, most eyepiece projection kits have several hollow tubes of various lengths so that you can set the effective magnification to just about any value you want. After experimenting a little, you will find that extremely high magnifications are more difficult to focus, are much fainter in the viewfinder, are more susceptible to

---

[181] The eyepiece projection tube has been called a tele-extender by some astronomers. This is an unfortunate nomenclature since the term also refers to a negative lens mounted in front of the camera which doubles the effective focal length of the telescope.

[182] If your camera does not have a removable lens then eyepiece projection, focused at infinity, may be used to project the image into the camera. While the system works in theory, the extra glass elements and possible internal reflections in the lens degrade performance by absorbing too much light. It is often simpler to buy a used camera which has a removable lens.

small vibrations in the telescope and may actually be blurred by atmospheric seeing. These problems are just more of the trade-offs which you will continually make regarding resolution, speed, ease of picture-taking and exposure time.

Once you have a good grasp on eyepiece projection photography, it is easy to make the transition to taking pictures of the planets. The exposure times will be slightly longer but the principles will be the same. Try Jupiter and Saturn first, for they have larger images. Mars shows polar ice caps and changing weather but it is much smaller in angular size even though it is usually closer to us. Venus shows phases, much like the Moon, but not much detail, as it is perpetually shrouded in clouds.

One can, at this point, start experimenting with color filters to enhance the contrast of transient features on the planets.[183] Moving your telescope from site to site in order to obtain the best atmospheric steadiness and resolution can also become a time-consuming challenge. I know several professional and amateur astronomers who never progressed beyond this stage but became masters of planetary photography, virtually ignoring the rest of the Universe beyond Pluto.

Deep-sky photography

Once you have photographed the Moon and brighter planets, it is time to search out among the stars. There are still some interesting objects within the Solar system such as comets and asteroids but we'll come back to them later after the basics of deep-sky photography are mastered. There are two great differences in technique when working with deep-sky objects. First, the exposures are tens of minutes to hours and, second, more pieces of equipment are needed.

No matter how well aligned your telescope is with the pole and no matter how much money you've spent on a telescope

[183] A word of caution is in order if you use print film or have slides reproduced. If the processing is done commercially, an automatic machine will try to make the total amount of light in your print equal to some ideal average for terrestrial photography. The machine will thus make the background of your astrophotos a washed-out gray, decreasing the contrast of the stars or planets you photographed. The only cure for this is to find a film processor who uses manual methods or contract for custom printing, which is often very expensive.

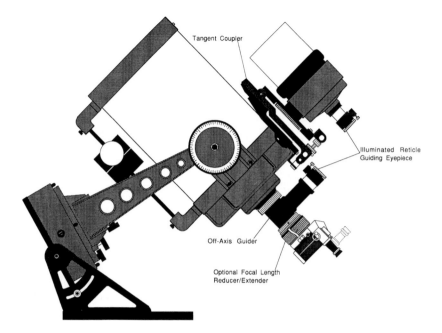

Figure 9.3. Accessories for guiding astrophotos.

drive system, there will be small tracking errors as your telescope follows objects across the sky. During an hour-long
astrophoto exposure, these errors will make the image of the
object of interest move slightly with respect to the image plane of
the telescope. Thus, you need some way of detecting these errors
and correcting for them.[184] The commonest method of correcting
the telescope pointing is to install an auxiliary optical system
which allows the observer to monitor the position of some star
within or near the field of view and issue commands to the drive
system which then makes the required adjustment.

Figure 9.3 shows the two most popular methods of guiding
equipment.[185] The T-shaped tube, called an off-axis guider,
attached between the main telescope and the camera contains a
small mirror or prism which steals light from the main optics and

---

[184] Even if you could make perfect gears for a theoretically perfect telescope drive, you
would still have to manually track the stars to correct for differential atmospheric
refraction which displaces stars slightly as a function of their elevation above the horizon.

[185] Several manufacturers produce small television cameras interfaced to the drive corrector which automatically keep the star centered. These are fairly expensive.

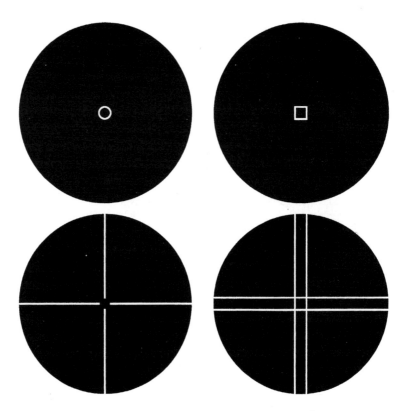

Figure 9.4. Typical guiding eyepiece reticles.

directs it up to an eyepiece. The mirror does not prevent light from the telescope reaching the camera since it images light adjacent to the field of view of interest. The eyepiece has a short focal length, typically six to twelve millimeters which yields fairly high magnification so that small errors can be detected. It has been suggested that the guiding eyepiece power should be at least 2.5 times the focal length of the telescope in centimeters.[186] Thus for prime focus photography, the 203.2-cm focal length implies a power of 508. I've used slightly less than this, about 330 power, without too much trouble. The eyepiece must have an illuminated reticle so that the observer can detect small errors in tracking. The reticle can have any one of several illuminated patterns projected on the star field. The common cross-hair should be avoided since the star can disappear behind the lines at the

[186] *The Sky: a user's guide*, David H. Levy, Cambridge University Press, 1991, p. 231.

center. An illuminated circle, box, open cross-hair or pairs of parallel lines perpendicular to each other will suffice.

The second method of guiding an astrophoto uses a smaller telescope mounted above the main optics. Note that this telescope has a much longer focal length and smaller field of view than the finder scope. It is important that this guiding telescope be rigidly attached to the telescope mount. Any flexure between the main optics and the guiding optics will show up as a tracking error.

Of the two approaches, most people use the off-axis guider because it is simpler and less expensive. The main drawback is that there must be a guide star bright enough to see in the off-axis guider which is adjacent to the object of interest. For objects in the thickly populated Milky Way this is not a problem but several external galaxies are considered difficult photographic objects simply because there isn't a handy guide star nearby. For these objects a separate guiding telescope is best. Some observers attach the guiding telescope to the main telescope tube using a tangent coupler which allows the guiding telescope to be accurately pointed up to five degrees away from the field of view of the main optics. While the tangent coupler is an expensive accessory, I have found it to be invaluable for astrophotography as well as in several other uses such as television astronomy.

The guiding telescope used by most observers is a long-focus refractor, usually a cheap telescope and often the first one the observer bought before he discovered the superior qualities of a Schmidt–Cassegrain. The older telescope is thus dusted off and pressed into service as a guiding telescope. I have found, however, that the length of refractors makes them more subject to vibrations when touched. At the suggestion of several friends I purchased a Maksutov telescope which uses mirrors and is similar in design to the Schmidt–Cassegrain. The telescope is physically shorter and thus vibrates less while it has the same effective focal length as a telescope 1 m long.

While the guide star images seen in the Maksutov aren't as bright as those seen in an off-axis guider due to the smaller Maksutov aperture, I can usually find a bright star within five degrees of any deep-sky object of interest. I have found, however, that maintenance on the tangent coupler is important. Each and every screw and bearing on this complex two-axis mount must be kept tight. Similarly, the Maksutov telescope and its fittings must be mechanically rigid. Failure to monitor these items generally results in flexure between the main telescope and the guider, resulting in tracking errors. While I prefer the versatility of using a separate guide scope and tangent coupler, I find that the process is much more finicky and requires more work than using an off-axis guider. Thus, for objects with easy nearby guide stars I'll use the simpler off-axis guider and only for difficult scenes will I use the tangent coupler.

Before attempting your first deep-sky photograph, align the telescope as accurately as you can to the Celestial Pole. Long-exposure photographs require exacting alignment or the observer will wind up making drive corrections constantly. Next, place the declination drive tangent arm so that it is in the middle of its travel. Celestial Pole misalignment may cause you to make declinations adjustments and you won't know whether you'll need North or South travel until after you've begun the exposure. Allow yourself adequate travel in either direction.

Now point the telescope at one of the brighter deep-sky objects such as the Orion Nebula, Andromeda Nebula (galaxy) or Omega Centauri,. The Magellanic Clouds and rich star fields of the Milky Way can also provide good opportunities.[187] Most objects will appear fainter in the viewfinder of a camera than they did through an eyepiece. This is because most cameras have a ground glass viewfinder which absorbs some of the light. Some camera manufacturers offer, as an accessory, a clearer focusing screen but before you purchase one, try it out. Many observers, especially those who are nearsighted, find using clear focusing screens difficult.

When focusing your telescope, always make the last turn of the focusing knob act to push the primary mirror away from the ground. The mirror support slides on a cylinder attached to the backplate of the telescope. If there is any play between the support and the cylinder then at some time during the exposure, the

---

[187] One possible glitch prevalent in older telescopes is caused by the possibility that the primary mirror central support may be a little loose with respect to the cylinder on which it slides to focus the telescope. Normally a little mechanical play between these parts is no problem but just after the telescope passes your central meridian, the mirror may flop from one side to the other of its travel. A 10 or 20 arc second shift is not unusual. This instantaneous shift in field of view can ruin a long exposure. If you have this problem, the cure is to plan your photos so the telescope never passes through the meridian during the exposure.

mirror may shift, causing it to point suddenly in a slightly different direction, ruining a long exposure. This is commonly referred to as mirror flop and it is not a design defect, for there must be some clearance between the support and the cylinder, else it would sieze up on cold nights or if dust entered the telescope. You will have to determine on your own telescope whether turning the focus knob clockwise or counterclockwise pushes the mirror away from the ground.[188]

Now find a guide star. You can rotate the off-axis guider around the main optical axis in order to locate an adjacent star which is bright enough. Once you have found it, however, ask yourself if the angle of the guiding eyepiece with respect to your observing chair is comfortable. You're going to spend the next half hour or so sitting still, staring into the guiding eyepiece while making small drive speed corrections. Practice guiding for a few minutes before you open the shutter. The main lesson here is to get comfortable first. Then take your pictures.

Advanced exercises

Comets and asteroids make fascinating photographs, especially because they are dynamic objects, not only moving through the Solar system but also changing in brightness and shape. Generally their faintness demands exposures of tens of minutes to hours. Unfortunately, over such a span of time they move with respect to the stars and thus their images are spread out over the picture. For an asteroid, this means that it shows up as a thin line amongst stellar dots. Faint asteroids may not show up at all since their image is spread out over a line.

One cure for this problem is to track the asteroid or comet and not the stars. In order to do this, an image of the object of interest must appear in the guiding eyepiece. For obvious reasons, an off-axis guider cannot be used and thus a separate guiding telescope

---

[188] Telescopes in the Northern Hemisphere turn counter clockwise when viewed from the tube end of the polar axis and the focus knob is usually on the right when viewing on the meridian at the Celestial Equator. If the last focus adjustment moves the mirror away from the ground then mirror flop will be minimized even when the telescope travels across the meridian while tracking. This is not so in the Southern Hemisphere, where the telescope turns clockwise. Some Southern observers have flipped the tube over so that the focus knob is on the left and thus prevents mirror flop when tracking through the meridian.

Figure 9.5. Film negative cut at wrong location.

is employed. In such astrophotos, the stars will appear as streaks but the comet or asteroid will not be smeared out by its motion.

Another advanced exercise is to catch two or more objects in the same field of view. Quite a bit of planning is required to catch a comet or asteroid in front of some familiar Messier Catalogue object. Publications devoted to deep-sky astrophotography often announce such conjunctions in advance. Chris Schur's photograph of Mars in the same field as the Pleiades demonstrates the principle in Fig. 4.7.[189]

Advanced mistakes

Time and experience will teach you new and different errors in astrophotography. A typical example is in Fig. 9.5 showing the Orion Nebula (M42) on both the left and right sides of the same frame. The automatic film slicer used by commercial film devel-

---

[189] The photo was taken by a 20.32-cm Schmidt Camera, which is slightly different than a 20-cm S–C in that the secondary is removed and a sheet of film is placed at the prime focus as shown in Fig. 1.2. Obviously, an eyepiece cannot be placed at the image plane since the observer's head would obscure the light path so the Schmidt Camera is a photographic instrument only. Some early Schmidt telescopes came with two front ends; one with the corrector plate and secondary Cassegrain mirror. The other, slightly longer front end used a corrector plate and prime-focus film holder.

opers looks for the first frame in the roll of slides, makes a slice and then counts sprocket holes to the next frame. If the first image on the roll doesn't have a clearly defined dark edge then the slicer picks a dark area (easy to find on most astrophotos) and slices there. The whole roll of slides wound up like this. The cure, of course, is to take one light-colored exposure at the beginning of the roll to set the slicer.

One of the most obvious mistakes is forgetting to set the exposure control before taking the first picture of the night. Nearly everybody at some time takes a 1/250 of a second exposure of some faint object followed by an hour and a half of practice guiding. The first clue you have that something is wrong is when you release the shutter control and the camera doesn't make a 'click' sound as the shutter closes. Another common error is to sit in an uncomfortable position. Half an hour into the exposure, your leg is cramped and you decide to move it. That's when you kick the tripod, slewing the object of interest entirely out of the field of view. Then there's the no-film-in-the-camera trick. You are free to invent your own new and unique mistakes. Your observing log is an ideal place to record such non-productive procedures. While you'll probably never forget the lessons you've learned, others who follow in your footsteps would like to avoid the pitfalls you've found.

While you are making a long exposure, time will seem to slow down, for guiding an astrophoto with a well-aligned mount is basically boring. Only a little correction in right ascension is needed to compensate for the final gear error in the drive train. If significant declination corrections are required then perhaps the polar axis isn't well enough aligned with the Celestial Pole. Meanwhile, your mind tends to wander, thinking of other things. The next thing you know, the guide star has drifted well away from the center of the field of view. Any bright stars in the photograph will now have a neat little line attesting to your lack of attention.

One trick that has been used successfully is to occupy your mind with non-visual exercises. Some people listen to music. I know one astronomer who composes particularly nasty limericks. He has a very interesting observing log. Several engineers, myself included, have occupied their time thinking of ways to avoid the boring job of guiding astrophotos while still enjoying the pictures. Such daydreams led me to sketch the design of a quadrant photodetector and servo-amplifier which I would interface with my drive corrector. Before I could finish final adjustments on the system, somebody showed me an advertisement for one of the first inexpensive solid-state TV cameras and I was off on a new project. Alas, my astrophotography activities suffered for lack of interest and I find myself currently more involved with electronic signal-to-noise ratios than photographic contrast transfer functions.

10

Photometers, computers, image intensifiers and television

Photometers

As the name implies, photometers measure light intensity. Modern instruments are much more sensitive than the light meter in your snap-shot camera. Indeed, some of the better photometers actually count individual photons. In astronomy, the photometer has become an invaluable professional tool which allows precise quantitative measurements to be made. Thus, the eye is no longer used for estimates of star brightness, eliminating all of the physiological problems of eyesight. This does not mean, however, that the problems such as variation of sensitivity with wavelength have been eliminated. The photometer has its own set of unique problems but they are much more amenable to analysis and calibration than the human eye.

There are two basic types of photometers, the photomultiplier tube and the solid-state photodiode. The photomultiplier, which has been around for half a century, is gradually being replaced by solid-state devices which are smaller, require less exotic high-voltage power supplies and are now more sensitive. You may elect to buy a complete photometer or build one yourself, as shown in Fig. 10.3.[190]

While a photometer measures all of the light falling on its detector, we generally don't want all of the photons entering the telescope tube to hit the light-sensitive element. A photometer placed at the image plane of your 20-cm S–C should be adjusted so that it sees the light from only a small area, typically the light from just one star.[191] Thus, all of the other stars in the field of view will not be measured. The aperture which blocks all

[190] For a description of the design and construction of photometers, see *Zen and the Art of Photoelectric Photometry*, Jeff Hopkins, HPO DeskTop Publishing, 1990; and *The Photoelectric Photometry Handbook, Volumes I and II*, David R. Genet, Russell M. Genet and Karen A. Genet, Fairborn Press, 1987 and 1989.

[191] The question has arisen often as to whether the 20-cm S–C is the proper tool for photometry since it has a corrector plate which may block some wavelengths of light. At

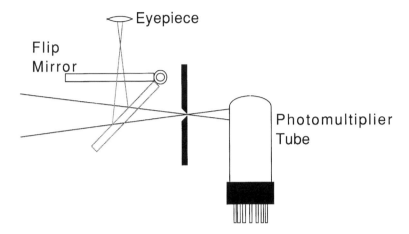

Figure 10.1. Simple photometer. This example uses a photomultiplier tube. The high-voltage power supplies and amplifiers are not shown.

of the starlight except that of interest should be small enough so that stars adjacent to the target star won't be measured but large enough so that all of the light from the target star will reach the detector element. The effects of seeing and resolution of the telescope must be taken into account here. Typically, apertures of a few arc seconds are used. Many photometers have several aperture holes of varying sizes, as shown in Fig. 10.2, mounted in a wheel so that the observer can select the optimum size for a given star and atmospheric seeing conditions.

When the starlight falls on the detector element and data are taken, light from the sky background will also contribute to the measurement. Thus, a complete observation includes a second reading of the blank sky adjacent to the star. Subtracting the second measurement from the first yields the brightness of the star.

least two manufacturers have modified the corrector plate with special anti-reflection coatings which optimize viewing in the visual band. My own 20-cm S–C is an older model without the coatings and I can see the difference visually as an improvement in the newer models. On the other hand, some of the coatings block the ultraviolet and infrared light. I have tested my intensified CCD camera on several telescopes and find that the infrared response is better by a factor of two on my uncoated telescope. Similarly, Jeff Hopkins, an astronomer who does multicolor photometric observations, reports that he can make publishable ultraviolet obsevations on his older model. The manufacturers of the telescopes say that the coatings can be stripped chemically from the corrector plate but that this will degrade visual band performance considerably.

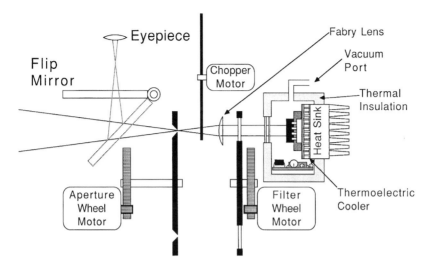

Figure 10.2. Complex photometer.

The output of the photometer may be a pointer on a current meter or a digital display showing the response of the detector. Some observers have interfaced their photometers with desk-top computers so that the data are logged automatically. In general, however, the data are not displayed in visual magnitudes but rather in microamperes of signal strength or photon counts. The effects of the detector's variation in response as a function of wavelength, the absorbtion of the atmosphere and the transmission of the telescope optics must all be factored into the data to determine the actual brightness of the star in visual magnitudes. While it is theoretically possible to determine the star's brightness by knowing the raw data and a calibration curve of detector response versus color, the detector may change its response with temperature too. Measuring how much starlight was absorbed by the atmosphere makes the job exceedingly complex, although not quite impossible.

An alternative method is to compare the apparent brightness of the star in question with one or more standard stars whose brightness has already been measured accurately. This is called differential photometry. Thousands of standard stars and their brightnesses can be found in several catalogues and reference works.[192] The method of observation is to record the brightness of

---

[192] I have found the most useful tables and finder charts to be those by Arlo U. Landolt, published in *The Astronomical Journal*, Volume 78, No. 9, November, 1973, p. 959.

the target star, record its background and then slew a short distance to a standard star. The slewing distance should be minimized so that the effects of atmospheric absorption, a function of the telescope's elevation above the horizon, are about the same for the target star and the calibration star. The brightness of the calibration star and its background are then measured and compared with the target star. Often a third star or check star is measured just to assure that the data are reasonable and consistent with each other.[193]

By this method the effects of the atmosphere, transmission of the telescope and several other difficult-to-measure parameters can be bypassed and the brightness of the star relative to the standard star can be calculated. This technique is best employed when the standard star is just about as bright as the target star or at least within three or four visual magnitudes of the star being measured. The reason for this is that the photometer, like all measurement devices, gives an output signal which is linearly proportional to the amount of input (light, in this case) only over a certain range. If either the target or comparison star is too faint then its measured value will be lost in the small-scale electronic noise of the instrument and the ratio of the two measurements will be inaccurate. If one of the stars is too bright then any one of several effects (including damaging the detector) may make the output data nonlinear.

Of course, astronomers want to measure fainter and fainter stars, groping for ever more distant objects. Thus, the effects of small signals become important and designers spend many hours concocting sophisticated electronic measurement systems that do not add noise to an already difficult measurement problem. For both photomultipliers and solid-state detectors, cooling the detector has been found to decrease the random thermal noise of electrons in the light-sensitive element. While cooling the detector decreases the noise, it increases the cost and complexity of the photometer, as shown in Fig. 10.2. Generally temperatures ranging from –40°C to that which can liquefy gasses are employed. The detector must thus be insulated from the atmosphere or water vapor will condense on its sensitive parts.

---

[193] For a discussion of photometric data reduction, see *Workbook for Astronomical Photoelectric Photometry*, Jeff Hopkins, HPO DeskTop Publishing, 1990.

Often this means sealing the detector in an inert atmosphere such as dry nitrogen. For those photometers using cryogenic temperatures, the detector must be sealed in a vacuum. Enclosing the detector in thermal insulation or vacuum implies that a method of getting the light into the detector must be engineered. Early detectors used a thermal window but later versions have employed fiber optics.

One further method of overcoming unwanted electrical noise is to convert the incoming light into a series of optical pulses. This is done with a chopper wheel which looks like a flat fan, as shown in Fig. 10.2. The blades of the chopper alternately block and pass the light. The signal output from the detector is thus a series of on-off-on-off pulses representing the light input. Any signal present in both the 'on' and the 'off' portions of the data must be due to electrical noise within the detector and thus it can be calibrated out of the final data. A further refinement of using chopped light is to employ a special electrical amplifier which responds to only a single frequency of input and rejects any other frequencies. If the light is chopped at this frequency then much of the noise (which occurs at all frequencies) will be blocked and only the signal of interest plus a small amount of noise which happens to be at the frequency of interest will be amplified.

Astronomers want to know not only how bright a star is but they also wish to know its color, for this tells us the temperature of the star, an important parameter in understanding how stars work. The color of a star can be determined by measuring its brightness through a set of standard colored filters. Generally, a set of filters transmitting in the ultraviolet, blue, visual, red and infrared are employed. The filters are usually mounted on a wheel or slide which allows them to be placed in the light path one at a time.

One final element in the photometer is the Fabry lens which defocuses any light coming through the aperture so that it is spread out over the entire sensitive area of the detector. The reason for this is that although detector manufacturers try to achieve a uniform sensitivity over the area of their detectors, one small region may be slightly more sensitive than another or there may be a smudge of dirt on the detector faceplate. Inserting a Fabry lens into the optical path makes the measurement insensitive to small

Figure 10.3. Typical photometric installation. Note the permanent mount, dew shield and large finders. Photo courtesy of Jeff Hopkins.[194]

pointing errors (within the field of view of the pinhole aperture).

All of the filters, pinhole apertures, cooling, choppers and other paraphernalia crammed into a modern photometer have the objective of reducing electrical noise and extraneous light which will degrade the measurement of photons from the star of interest. One final noise source is the sky, for seeing effects will show up in the data. From moment to moment the star will appear brighter and fainter. One obvious method of overcoming this is to measure the star many times per second and average the result. This, however, takes time and ideally astronomers would like to make hundreds of measurements per night. The observer must decide whether more minutes should be spent measuring a star in order to obtain yet another fraction of a per-cent of precision. At some time the point of diminishing returns is reached and it is time to press on to the next measurement. Using today's detectors, only a few seconds of measurement will generally yield precisions of a small fraction of a visual magnitude.

[194] The photometer and observatory, both built by the observer, include a computer inter-face which records individual photon counts and stores the data on disk. While Jeff has tested several commercial photometers, he seems to prefer his own designs.

The complexity of a modern high-precision photometer may appear daunting to some but for others it is a wonderful design challenge involving the interaction of mechanical engineering, physics, optics, thermodynamics and electrical engineering. Indeed, there are astronomers who have so much fun inventing instruments that they now consider their 20-cm S–C telescope merely a test bed for their next creation. These folk seldom observe other than to test a new instrument or demonstrate it to potential customers. They are as necessary to astronomy as observers, for they bring us the new and improved instruments to peer farther into the Universe.

One might ask what objects could be observed and what information gained through photometry. Most stars appear to shine at a constant brightness, seeming never to vary other than when the seeing is bad and then they just twinkle a bit more. Astronomers have found, however, in their search for standard calibration stars that almost all stars vary in brightness just a little. Even our own Sun experiences periods of minute fluctuations in brightness. Fortunately, most stars are constant in brightness to the accuracy of a few thousandths of a visual magnitude. For most photometric work, we can treat these stars as unvarying standards.

There are thousands of stars bright enough to measure with a 20-cm S–C telescope which vary regularly in light output. The American Association of Variable Star Observers (AAVSO) and other similar organizations world wide have organized amateurs and professionals into networks of teams to study these stars. Thousands of other stars are suspected of varying but have not been examined closely enough. There are not enough telescopes in the hands of professionals to observe all of the suspected variables and thus amateurs have contributed greatly in this area.

Most of the stars of interest to AAVSO vary in brightness from week to week and thus do not require continual monitoring. Such long-period variables can be observed a few times a month and their fluctuations in visual magnitude noted over the course of a year or more. Once all of the data are collected, the shape and magnitude of variation over time may tell an experienced observer exactly what mechanism is causing the star to change brightness. On the other hand, some unusual light curves have caused both professional and amateur astronomers to scratch their heads in wonderment.

In a typical variable-star observing session the astronomer first prepares an observing plan, detailing which stars will be observed. The plan should include finder charts for the stars of interest, the comparison stars and the check stars. In order to minimize the effects of the atmosphere, stars are observed as high in the sky as possible. Thus, stars in the West will be observed first before they set. After determining that the sky is sufficiently clear and steady for reliable observations, the telescope is slewed to the first star of the night. A reading is taken and then the adjacent background sky brightness is measured. Usually only the raw data are recorded and the two readings are subtracted later, after all measurements have been completed. Some observers, however, prefer to reduce the data immediately, especially if they have a computer handy in order to spot observing or instrument problems before they have wasted a whole night with a malfunctioning instrument. After measuring the star of interest and its background, the telescope is slewed to the comparison star and the process repeated. Then the check star is measured. The entire sequence may also be repeated for a different color filter. When observing the three stars and their backgrounds in six colors, thirty-six separate readings will be noted. This constitutes one complete set of observations on one star of unknown brightness.

In order to handle the large volume of numerical data accurately, many observers have taken to entering the raw readings directly into a desk-top computer in order to record and process the information. The observations can be time consuming but the reward of seeing an accurate light curve unfold before your eyes is well worth the effort. After many observing sessions, the secrets of the stars will begin to appear on the graph paper or printouts of your reduced data. There is also a certain satisfaction to seeing your data used by other astronomers in order to fathom the nature of the stars.

Photometry can be used for observations of other objects, most notably comets and asteroids. Since these objects move around the sky, finding them from night to night is a bit more involved. Finder charts for objects of interest can be obtained from the Association of Lunar and Planetary Observers (ALPO).

Asteroids are more rewarding in that they change their brightness much more rapidly than most stars. Asteroids are irregularly shaped chunks of rock and most rotate with a period of just a few hours. The Sun may illuminate the broad side of an asteroid and then just a few hours later, shine on the small end of the same rock. Usually, within a single night an entire rotation can be recorded. Asteroids can also exhibit effects due to the relative Sun–asteroid–Earth phase angle. While most observations are made at opposition when the asteroid is more or less opposite the Sun in the sky and thus we view the fully illuminated side of the rock, observations made when the Sun is off to one side can yield information as to the shape of the rock. There is some speculation that many asteroids have dark or light spots which will effectively change the brightness of the reflected Sunlight as the rock rotates.

Comets pose a slightly more difficult problem for photometrists in that they are not point sources like stars and asteroids. If the pinhole aperture of the instrument is opened up to include light from both the nucleus and the coma then a large signal will be included from the sky background and while this may be subtracted later, the resulting difference is usually so small that it is of the order of magnitude of the noise (sky and instrument errors) of the measurement. Consequently, most observations are of the brightness of the nucleus only. Comets have been known, however, to show variations in brightness over periods of minutes and thus they can be dynamic and interesting objects. Cometary photometry can also include the study of polarized light in order to look at the interaction of the gas and dust with the Solar wind. Thus, the filter wheel may contain one or more polarizing filters or a separate rotatable polarizing filter may be added to the optical train of the photometer.

Eclipses and mutual occultations of the Galilean satellites of Jupiter, discussed earlier under visual observations, are a suitable target for photometry. The time of an event can be determined by recording the sometimes subtle changes in brightness as one satellite slips into the shadow of another or passes behind another. Typically these events take minutes to complete and happen every few weeks. While it is at times difficult to make precision satellite brightness measurements in the glare of nearby Jupiter, the results can be valuable. One requirement for

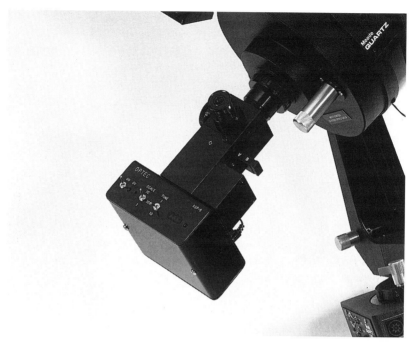

Figure 10.4. Commercially produced photometer including built-in filter slide and photon-counting electronics. Photo courtesy of Optec Incorporated.

such observations is that the time of observation be recorded much more accurately than for stellar, asteroid or cometary photometry observations. While a one-minute accuracy in time may suffice for other measurements, timing with an error of about one second is required for Jovian events.

Time-resolved photometry implies that some method of recording events be employed. Various systems using stereo audio tape recorders with one channel handling the photometric data and the other recording time have been developed. Of course, data-reduction electronics must also be developed in order to reduce the data. Often computers with internal clocks are used for these measurements.

An even higher degree of timing precision is required for occultation timings, typically around 0.01 to 0.001 seconds. In these events a star is occulted by either the Moon or an asteroid. Occasionally planetary or cometary occultations are observed but these are rare.[195] Lunar occultations can tell us several things about the star and the Moon. As the star winks out by passing behind the dark edge of the Moon, the time of the event fixes the Moon's position with respect to the stars precisely, thus aiding in calculating the Moon's orbit.

The star may not wink out in one instant. Indeed, it may fade slowly, indicating that it is a large and not quite point-source object. The diameters of several nearby giant stars have been verified by this method. More interesting, however, is the type of star which suddenly decreases to a fraction of its usual brightness. Then a split second later it disappears altogether. This is a sign that the star is a double whose components are so close that they cannot be split with even the highest-resolution telescopes. Many previously unsuspected double stars have been discovered this way. While visual Lunar occultation observations have been used for years to discover new double stars, a high-speed photometer with millisecond time resolution can discern those events which happen too quickly for the human eye to notice.

Spectroscopy

Spectroscopy involves breaking the light from an astronomical object into its various colors and studying the relative brightness of those colors. Many facts can be discerned from starlight in this manner. For instance, the temperature of the star can be calculated since hotter stars emit more blue light and cooler stars have more red. Some stars and nebulae shine strongly at one or two characteristic wavelengths, indicating the presence of some particular type of atom or element.[196] Others show a dark line at characteristic wavelengths indicating that something in the star or interstellar medium is absorbing at that wavelength.

Spectrographs are often large, bulky things, sometimes rivaling the main telescope tube assembly in size and mass. The light

---

[195] The faint, dark rings around the planet Uranus were first observed by watching a star pass behind the planet in order to measure its atmospheric density. Just before and just after the event, brief occultations by the rings were recorded, thus signalling a new and unexpected discovery.

[196] Each atomic element emits and absorbs light at specific wavelengths unlike any other type of element. This "fingerprint" allows us to identify chemical elements in astronomical objects.

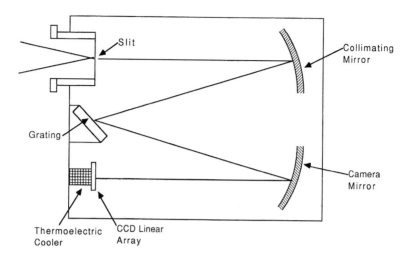

Figure 10.5. Czerny–Turner Spectrograph.

is split into its colors using either a prism or a ruled grating and then imaged onto a piece of photographic film. Spectrographs are characterized by their dispersion, measured in ångströms per millimeter.[197] This tells how many different colors are imaged onto each millimeter of film. High-dispersion spectrographs spread the light out so that differences in adjacent colors can be discerned easily. Unfortunately, this also means that not many photons reach each section of film and thus exposures must be longer to compensate.

In modern spectrographs the film has been replaced by a CCD (Charge Coupled Device) linear array composed of a row of several thousand photodiode detectors. Each detector samples a small portion of the spectrum. After integrating the starlight for an appropriate time, the brightness at each individual photodiode can then be sampled and read into a computer. The data can then be plotted as a graph for study or classification by the astronomer.

The spectrograph shown in Fig. 10.5 uses a replica grating with 1200 lines per millimeter to spread the light to a dispersion of about 63 Å/mm. It is of the Czerny–Turner configuration in that it uses a small mirror to collimate the light from the telescope and a second small mirror to project the spectrum onto the

---

[197] Ångstroms are the usual unit measuring the color of light. Blue light has a wavelength of about 4400 Å while deep red light occurs at about 6000 Å.

Figure 10.6. Small spectrograph mounted on a Schmidt–Cassegrain.
Photo courtesy of Andrew Grunke, © Aspect Instruments.

array. The 5000 photodiodes in the array are on 7 micron centers so each diode samples the light from only 0.44 Å of the spectrum. The entire spectrograph, mounted in a sealed box with a thermo-electric cooler weighs less than 2 kg.

With this instrument, stellar classification of stars can be performed. It can also be used to study the changes in spectrum over time of some variable stars. Many of these stars change brightness periodically in regular cycles but others flare unexpectedly. Such events are often first noticed by amateurs simply because there are so many more of them looking at the sky. With relatively inexpensive spectrographs such as this available, it will be possible to obtain spectra of flare stars very early in their outbursts.

Television

With the advent of inexpensive television cameras, the possibility exists to record your observations without having to resort to sketching or photography. TV also has the added advantage that several people can watch the screen at once, making group

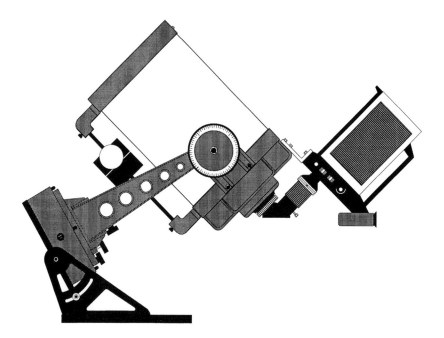

Figure 10.7. Large television camera mounting. The camera, using early 1980s technology, is about as large as a current camcorder. Note that a separate support bracket and extra counterweights are needed.

observing sessions considerably more interesting. There are two distinct types of TV sensors for the small telescope.[198] The first is an adaptation of a conventional home video camera.[199] These cameras run at standard TV rates and interface easily with conventional video tape recorders and monitors. They run at 25 or 30 frames per second and thus for each frame, they collect only a few photons. The second type of camera integrates photons in a time exposure and, after several seconds or minutes, reads the summed image out to a display or computer. These cameras are much more sensitive but require a special displays and video storage media to see results.

---

[198] See the two-part article titled *Television in Amateur Astronomy* in ASTRONOMY *Magazine*, November, 1984, p. 50, and December, 1984, p. 35.

[199] It is best if the objective lens of the camera can be removed so that the telescope's image is projected directly onto the sensitive element of the camera. If the lens is non-removable, then an eyepiece projection method similar to that discussed under photography may be employed.

We will initially consider the use of standard-rate TV cameras. These are available in both tube types and solid state versions, also called CCD cameras. The installation shown in Fig. 10.7 shows an older model tube-type TV camera mounted on a 20-cm S–C telescope. I had initially attached the camera straight out behind the mounting plate but found that the camera struck the tripod and fork arms in too many orientations. Thus, the folding mirror was added. This yields an inverted view of celestial objects but visual observers have been dealing with that problem for years. It took a bit of garage-shop level engineering to make the mounting and camera attachments stiff enough that the whole image didn't shake disconcertingly when the telescope was touched.

Typically, the active area of TV sensors is smaller than the usual 35-mm photographic camera frame and thus they show a smaller field of view. This is an advantage since, other than the Moon, almost all objects which are bright enough to be detected by a common TV camera are the brighter planets. Stars down to about Mv 4 can be seen but seldom do two or more stars appear in the same frame. This set-up has been used for demonstrations to classes of school children and with the local TV news teams when bright astronomical events occur. This method of observation has been helpful when our astronomy club scheduled a public star party at which a hundred or more guests showed up and all we could see was the bottom of a dense cloud layer. At least we could bring out some video tapes from earlier observations and give the guests a sense of what astronomy is like.[200]

In Fig. 10.7 it can be seen that the camera's viewfinder (actually a small TV screen) has been turned so that the observer can adjust the position of the telescope while watching the output of the camera. I have since found that it is easier to watch a small monitor placed beside the telescope rather than bend over to look into the viewfinder. I have taken this one step further in that once I have set up the telescope, I leave the area and go inside to watch a larger screen in comfort while operating remote controls

---

[200] This configuration has also been used in early tests to observe asteroid occultations in which the size and shape of the asteroid may be determined when the minor planet passes in front of a bright star, blocking out its light for a few seconds. The camera shown has since been replaced by a more sensitive one.

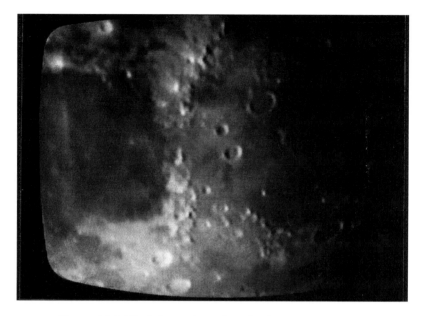

Figure 10.8. The Moon viewed with a home TV camera.

to point the telescope. This way I can avoid cold nights, hot evenings[201] and mosquitos.

An integrating TV camera works like a photograph in that the longer the exposure, the fainter the object recorded. The similarity ends there, however, in that the typical resolution of most CCD cameras today is nowhere near that of film. Typical CCD cameras have a resolution of several hundred picture elements (pixels) across the width of the frame. In addition, integrating cameras almost always operate in a black and white mode – they lose color information.[202]

CCD cameras are typically smaller than either camcorders or tube type TV cameras. This makes the mechanical interface much simpler. One characteristic of integrating CCD cameras which operate with exposures longer than about one second is

[201] Where I live in Arizona the temperature often exceeds 38°C (100°F) at midnight.

[202] Some observers have fitted their integrating CCD cameras with a rotating filter wheel which allows them to make separate exposures in the red, blue and green bands. The three pictures can then be summed in a computer image processing program to produce color photographs. Aligning the individual picture elements of the three pictures and balancing colors can be a tricky business, however.

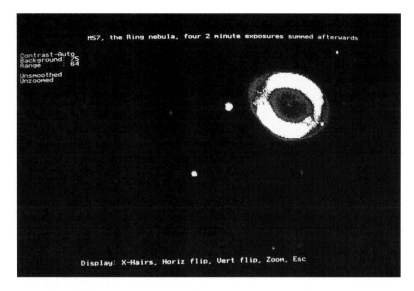

Figure 10.9. Ring Nebula (M57) CCD camera exposure. Prime-focus image from a 12.7-cm S–C telescope, four two-minute exposures summed later in a computer. Image by Joe Perry.

that they must be cooled internally to at least dry ice temperatures of about –40°C (–40°F). Most have built-in thermoelectric coolers and are filled with either a dry, inert gas or vacuum in order to prevent frost from forming on the cold sensor chip. In spite of some resolution and equipment cost problems, the results can be spectacular, as shown in Fig. 10.9.

Pretty pictures are not the only possible product of TV observations. The CCD camera chip, composed of thousands of individual photodiodes, has the potential of making exacting scientific measurements useful to organizations such as the American Association of Variable Star Observers (AAVSO). Using techniques pioneered by the International Amateur-Professional Photoelectric Photometry organization (IAPPP), the camera can be calibrated using standard stars and then used to measure variables in much the same way that photometers are used.[203] The data reduction and chip calibration are

---

[203] At the time of writing the AAVSO was engaged in a project to develop manuals and software for standardized CCD camera measurements of variable stars. By the time of publication, the project should be complete.

a bit more complex but software is now available for most common CCD cameras which allows the observer to accomplish this.

Image intensifiers

In order to perform occultation timing in which an asteroid or the limb of the Moon passes in front of a star, I needed a sensor which would record the time of an event to within a fraction of a second. Unfortunately, most common TV systems on a 20-cm S–C telescope can see stars down to only about Mv 4 with a 1/30-second integration time. Since most occultations involve much fainter stars, I either needed a larger telescope or some sort of light amplifier.

An image intensifier amplifies the brightness of the telescope's image and brings it up to a level which can be sensed by the standard scan rate TV camera. It is not without its defects, for it also intensifies the sky background brightness, any fluctuations in that brightness caused by seeing and adds its own amplifier noise to the image. The devices are also fairly expensive, although they are appearing on the military equipment surplus market now.[204] The intensifier must also be engineered into the optical train of the telescope, as shown in Fig. 10.10.

First, the field of view of the intensified TV camera was so small (about 1/3 of a degree diagonal) that I had to add a 'superfinder' on top of the telescope. The object of interest is first located in the six-power finder and moved to the center of the field of view of the finder scope. It is then located using the 100-power superfinder and moved to the center of that field of view. Usually, the object will then appear on the camera output.

A special mounting had to be made to hold the image intensifier and couple its input plane to the image plane of the telescope. In order to be able to use various effective focal lengths, the coupling thread was made to match that of photographic cameras so that tele-extenders and tele-compressors could be inserted in the beam. The intensifier requires a high-voltage power source but the amperage requirements are low so a com-

---

[204] Ben Mayer has quipped that an image intensifier magnifies an image by a factor of three thousand six hundred dollars.

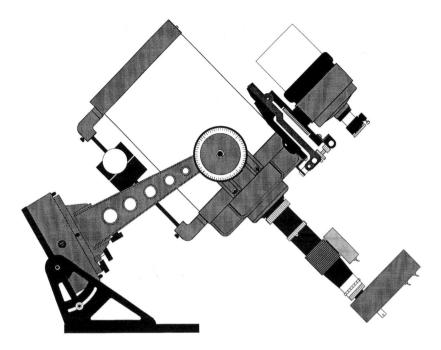

Figure 10.10. Image intensifier and CCD camera installation.

mercially made battery-operated power supply was built into the box attached to the intensifier. The output of the intensifier is a small glowing image, much like a miniature TV screen. Professional astronomers couple the output of the intensifier to the TV camera via fiber optics but these devices cost thousands of dollars for custom installations. I elected to use a small relay lens (actually, the 12-mm focal length objective which came with the TV camera) modified so that it would focus at a distance of about 1.5 cm in front of the lens. Thus, it acts like a macro lens for the TV camera, focusing on the output of the image intensifier. This design is horribly inefficient, losing some 90% of the light from the intensifier before it reaches the camera. Consider, however, that the intensifier has a gain of about 5000. The relay lens is about 10% efficient. Thus, the system gain is still about 500 which is not bad for something lashed together in a garage.

With the intensifier, I can see stars down to about Mv 12 which, for my standard rate TV camera, would require a meter-class

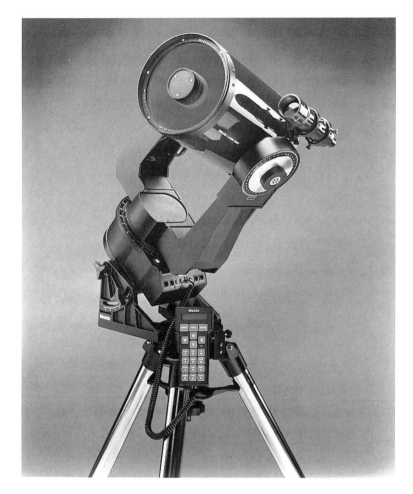

Figure 10.11. Computer-controlled telescope. Photo courtesy of Meade
Instruments Corporation.

aperture telescope without the intensifier. The intensifier is much
less expensive than purchasing a meter-class instrument and it is
also much more portable.

Computers

Computers seem to be appearing everywhere and astronomy is
no exception. In Chapter 2 the use of a computer as a planning or
note-taking device was discussed. Beginning in the early 1980s,

however, some avid computer experimenters started hitching home computers to 20-cm S–C telescopes. Initially, pulse-counting encoders or electronic setting circles were attached to the axles of the telescope so that the right ascension and declination could be displayed. Soon, motors of various types were added so that the computer could control the pointing of the telescope. In those days, some truly cumbersome mechanical engineering blunders were invented. Eventually, the technology became more refined and the commercial manufacturers now produce standard models which are controlled by microprocessors such as the one shown in Fig. 10.11.

The microprocessor-based computer in the hand paddle stores the locations of many objects in a standard catalogue plus any special coordinates which the observer would like to remember. With an instrument such as this, one can whiz through the Messier Marathon faster than the sky rotates. This may seem hardly sporting to those who painstakingly learned the positions of the objects and who participate in contests to see how rapidly and accurately they can acquire the nebulae and clusters. On the other hand, an observer who is making variable-star measurements would welcome semi-automated control of the telescope since it allows him to spend less time finding the star of interest and more time making useful measurements.

Some of the computer controlled instruments will even allow for sloppy alignment of the polar axis. The observer simply lines up several stars and tells the computer when the star is in the eyepiece. The computer can then calculate the misalignment and compensate for it in future pointing. Taking this to an extreme, one group has set up its telescope in an altitude–azimuth configuration and lets the computer point the telescope, allowing for the 57° misalignment of the polar axis. Since the telescope is used mainly for visual stargazing over short periods, tracking manually is not a problem.

As a final element of automation, some commercially produced computer-controlled telescopes allow input from the observer's desktop computer. This raises the possibility that computer controlled sensors such as photometers and CCD cameras can actively point the telescope through a feedback loop. This effectively removes the astronomer as an observer and places him in the position of observatory manager/planner and data reduction analyst.[205] Of course, the installation of a cloud or rain sensor might be in order if the telescope is to run unattended. Several observers have equipped larger S–C telescopes in this manner in order to study the light curves of long-period variable stars which need to be monitored once every few days. On any given night, the telescope may sample a hundred or more such stars and store the data away for analysis months or years later. There are many more interesting stars within the detection range of a 20-cm S–C than there are professional telescopes to go around.

A second application for unattended observation is monitoring stars known to suddenly brighten once every few years. These are called cataclysmic variables and when one goes off, professional observers with large ground and space-based telescopes want to be notified so that they can record this rare event. Several observatories are being set up with small telescopes to patrol the hundreds of suspected variables in this manner. This is one of the areas where amateurs can help professional astronomers.

---

[205] For a discussion of completely automated telescopes, see *Robotic Observatories*, Russell M. Genet and Donald S. Hayes, AutoScope Corporation, 1991, and *Robotic Observatories: Present and Future*, edited by Sallie Baliunas and John L. Richard, Fairborn Press, 1991.

11

Afterword

In this book I have tried to suggest some of the activities which may be pursued with your 20-cm S–C telescope. In doing so, I have necessarily skimped in some areas such as nature photography and sketching simply because there are so many different other observations which can be made with the instrument (and because frankly, I don't do much nature photography or sketching). There are probably thousands of uses which haven't even occurred to me.[206] If you do invent some new uses, please let me know. My address is in the heading to the index.

While writing this book I was asked by a colleague working on artificial intelligence computer systems if the book could be turned into an 'expert system' type of computer program. Such programs operate on a set of rules and can answer general questions about their area of expertise. I am unsure as to whether such a program can (or should) be written. In elaborating The Rules For Using Small Telescopes, however, I realize that part of the rules depend on what the observer wants out of the observation. Some folks want a hobby, a diversion, something to take their minds off of the job. Tough observing projects aren't for them. Some folks want to conquer something, make a 'personal best'. Them, we give a list of faint double stars to split. Others want to contribute to science and may bring to the observation some skill such as sketching or electronics knowledge. Those folks belong in the American Association of Variable Star Observers (AAVSO) or the Association of Lunar and Planetary Observers (ALPO) or the International Occultation Timing Association (IOTA).

---

[206] One observer, Ben Mayer, told me he had bought the telescope to observe a bikini-clad Hollywood starlet who lived down the hill from him. Only after she left the pool area and the Sun set did he think of other uses for the instrument. He has since become a confirmed amateur astronomer and author on the subject.

How you observe and what you observe depend on what you want out of astronomy (and some folks want to give something back to astronomy). The 'rules' may be broken freely and that's how we often learn of new techniques. But breaking the rules usually means that the observer sees less than an optimum view. As newcomers, some astronomers might want to bypass all of the 'easy' Messier objects as beneath their dignity and go straight for the difficult to find NGC objects. They are free to do so but I suspect that observing will be more difficult than it needs to be.

What you see through your telescope depends on what you bring to the eyepiece. If you step up to the instrument armed with a couple of star charts, some curiosity and an eagerness to learn then you will probably be astounded at the sometimes bizarre occupants of the Astronomical Zoo. If you come prepared to do battle with the Great Unknown and have a little technical training then you might actually push back the frontiers of Science – or at least nudge it a little. If all you want to do is let your mind escape the hum-drum world for a little while, then gaze into the glass and you can contemplate scenes at the edge of infinity.[207] It has been said that 'Joy in looking and comprehending is nature's most beautiful gift.'[208]

People observe for a variety of reasons. Some will tell you of the beauty of the Heavens. Others will rattle on about the great clockwork in the sky, marching onward inevitably. A few might mention reaching out and touching the face of God. Why do I observe the stars? Well, every once in a while when all of your planning goes just right and the equipment holds together and the sky stays clear and you just happen to be in the right place at the right time, then you get to snatch one little piece of information from Mother Nature – and it's something you never knew before – and THAT is what it's all about.

As an illustration, I will relate the circumstances of the occultation of the star SAO 146599 by the asteroid 47 Aglaja on 16 September, 1984, as shown by the pictures in Fig. 5.2. I was the

[207] Infinity is a tough concept to grasp for many people. As an undergraduate Physics student I was asked to wrestle with it in equations and not just contemplate it as an abstract concept. The great physicist Dr Peter Gabriel Bergmann gave his students some insight into the matter when he said, 'Infinity...' (we all scribbled in our notebooks furiously) '...is farther than half-way to St. Louis – and this works for any point in the Universe!' Upon reflection, I have never heard a better definition.

[208] *Ideas and Opinions*, Albert Einstein, Crown Books, 1955, p. 28.

expedition leader of half a dozen teams of amateur observers spread out over central and Southern Arizona. Lowell Observatory in Flagstaff, AZ, had also fielded several teams and we coordinated our geographic positions so that our data sets would compliment each other. I set up near Queen Creek, AZ, and erected my 20-cm S–C telescope as the Sun set. This was the first time I had used a new configuration of an image-intensified TV system so I allowed myself plenty of time to check and adjust the electronics. I entered the required notes in my observing log concerning weather, telescope configuration and the purpose of the observation.

We often arrange to stay in two-way radio contact when observing alone in the desert and this expedition was no exception. I chatted several times with the observers a few kilometers North and South of me as I tweaked the equipment into final focus. Twilight threatened to hamper acquiring the star and its low elevation near the Southeastern horizon did not present the best aspects. My observing log mentions curious streamers radiating from the East, probably a Sun shadow. By 18:47 hours local time (01:47 Universal Time) I could see the star Vega. The weather remained pristine, however, and soon I had an image on the TV monitor of SAO 146599. I'd practiced locating it in twilight several days earlier.

Hidden coyotes sang to the night sky as I waited in the desert. Shadows deepened among the cholla bushes and a distant saguaro cactus raised its arms like a hideous ghost. I calibrated and checked the video time inserter which is a key element in video occultations. I could not see the asteroid Aglaja on the monitor as it approached the star, for its magnitude was at the very limit of my system's sensitivity. A few minutes before the event, I started the video tape recorder rolling and made one last radio check with adjacent observing stations.

Everything ran perfectly. The main power battery held its voltage. The telescope tracked true, requiring only one minor adjustment. WWV radio time signals murmured to the video tape recorder. The twilight deepened to an inky black attainable only under the clear, dry skies of Arizona. The shadow of Aglaja rushed across the landscape and was caught by careful astronomers fully prepared to make the most of the observation.

For 12.31 seconds on that evening the star SAO 146599 was not

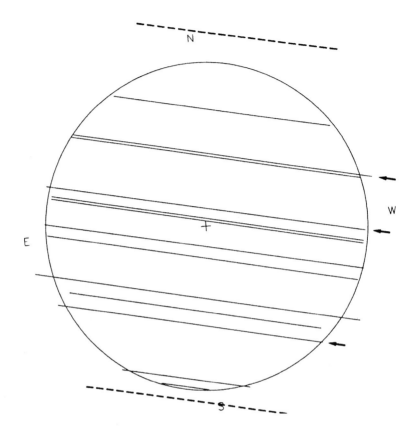

Figure 11.1. Reduced data showing the asteroid Aglaja. Illustration courtesy of Dr Robert Millis, Lowell Observatory, reprinted with permission of *Icarus*.[209] The arrows indicate photoelectric observations.

visible from Queen Creek, AZ. During that time I fancied I could see the asteroid itself as it covered the star. This was astronomy in its most intense meaning! Suddenly the star popped back into view and it was all over. I waited several minutes, recording the unwavering star in case any companion objects to the asteroid might lurk nearby. None were seen, almost an anticlimax to a memorable observation.

I grabbed the radio microphone and asked the nearby stations what they had seen. We had all observed occultations and in a

---

[209] *The Diameter, Shape, Albedo and Rotation of 47 Aglaja*, R.L. Millis, L.H. Wasserman, E. Bowell, A.W. Harris, J.W. Young, M.A. Barucci, R.M. Williamon, P.L. Manly, D.W. Dunham, R.W. Olson, W.E. Baggett and K.W. Zeigler, *Icarus* no. 81, 1989, p. 375.

few minutes they would report reduced data. Meanwhile, I packed up the telescope and unpacked a small portable computer. I had programmed the orbital elements of the asteroid into the computer and awaited input from my collaborators. Soon the others reported as to when the star disappeared and reappeared. I entered the data and waited breathlessly. Quickly, the characters on the screen glowed: the diameter of the asteroid Aglaja is 131 km, based on a first approximation ellipsoidal shape.[210]

I leaned against my van and grinned. Only one man on Earth knew the size of this asteroid. For the space of a dozen heartbeats I kept the secret and found the joy of discovery – a feeling I have seldom experienced except at the telescope. Were these the sensations that Galileo felt when he first viewed the Jupiter system? Did Newton glow like this when he finally realized why the Moon circles the Earth? Certainly, my meager discovery was a poor second-best to their monumental contributions – but in this case, second-best still felt outstanding.

Then I called my friends on the radio and told them the numbers.

As I packed up the last remnants of my equipment, I glanced at the sky and chuckled with hubris – the wanton insolence or arrogance resulting from excessive pride.[211]

---

[210] The actual diameter after reducing all data is 136.4 ± 1.2 km, reference the same paper as the previous footnote.

[211] *Webster's New Twentieth Century Dictionary, Unabridged Second Edition,* World Publishing Company, 1971, p. 883.

Appendix 1
Sources of further information

Books

These are the volumes on my bookshelf that are dog-eared and worn. They are right beside my desk so I don't have to get up when I need one. This is by no means an exhaustive list but rather a set of references which I've found handy. You may find others which are perfectly good too. I have given the ISBN identification of books wherever possible. The reader should note that often these standard reference works are updated with later editions or are made available in soft cover. Such changes generally alter the ISBN but the contents remain essentially the same.

Allen's Astrophysical Quantities, C. W. Allen, The Athlone Press, 1973,. ISBN 0 485 11150 0

Astronomy, Robert H. Baker, Van Nostrand Press, 1964

Astrophotography for the Amateur, Michael Covington, Cambridge University Press, ISBN 0 521 40984 5

Burnham's Celestial Handbook, Robert Burnham, Dover Publications, Inc. ISBNs for the clothbound series are: 0-486-24063-0, 0-486-24064-9 and 0-486-24065-7

The Cambridge Astronomy Guide, Bill Liller and Ben Mayer, Cambridge University Press, ISBN 0 521 39915 7

The Cambridge Deep-Sky Album, Jack Newton and Phillip Teece, Cambridge University Press, 1983

Introduction to Asteroids, Clifford Cunningham, Willmann-Bell, Inc., 1988, ISBN 0 943396 16 6

The Sky: A User's Guide, David H. Levy, Cambridge University Press, 1991, ISBN 0 521 39112 1

One set of books stands out, however, and that is the Harvard

214

Books on Astronomy. Through the years they have been updated but the original authors and titles have remained. They are

Earth, Moon and Planets, Fred L. Whipple
Between the Planets, Fletcher G. Watson
Our Sun, Donald Menzel
Stars In The Making, Cecilia Payne-Gaposchkin
Atoms, Stars and Nebulae, L.H. Aller
The Milky Way, Bart J. Bok and Priscilla F. Bok
Galaxies, Harlow Shapley
Tools of the Astronomer, G.R. Miczaika and William M. Sinton

Atlases and catalogues

Norton's Star Atlas,, 18th edn, Arthur P. Norton and J. Inglis, edited by I. Ridpath, Sky Publishing Corporation, 1989, ISBN 0 85248 900 5

The Sky Atlas 2000.0, Wil Tirion, Cambridge University Press, 1981, ISBN 0 521 24467 6

SAO Catalogue and *SAO Atlas* (for advanced observers)[212]
Skalnate Pleso Atlas of the Heavens

Periodicals

Most periodicals listed below are in the form of printed material but at least one is distributed electronically via computer. Electronic media are also used for radio and television programs dedicated to astronomy. Three worth mentioning are *Star Hustler*, a TV show on the Public Broadcasting System, *Star Date* on many syndicated radio stations and *The Sky at Night* on BBC Television.

Abrams Planetarium
Monthly Star Charts
Michigan State University
East Lansing, MI 48824,

---

[212] The *Smithsonian Astrophysical Observatory Star Catalogue* is a four volume set of books with 258 997 entries. It can also be obtained as a set of computer readable disks. The *SAO Atlas* is a collection of 152 star charts showing stars and nebulae as faint as Mv9.

ASTRONOMY Magazine
Kalmbach Publishing Co.
21027 Crossroads Circle
P.O. Box 1612
Waukesha, WI 53187, USA
414-796-8776, Fax: 414-796-0126
CompuServe 72000.2704

Journal of the Astronomical Society of the Atlantic
This is an electronic journal published on several computer bulletin boards and nets. See the section on computers below for information on how to access nets.

Astronomy Now
193 Uxbridge Road
London W12 9RA
United Kingdom

Sky & Telescope Magazine
Sky Publishing Corporation
P.O. Box 9111
Belmont, MA 02178-9111,
617-864-7360, Fax 617-864-6117, Hotline 617-497-4168
Internet: skytel@cfa.harvard.edu

Mercury, the Journal of the Astronomical Society of the Pacific
390 Ashton Ave.
San Francisco, CA 94122, USA
415-337-1100, hotline recording 415-337-1244

Yearly publications;

Observer's Handbook,[213] Roy Bishop; Sky Publishing Corporation

The Astronomical Almanac[214]
Available from;

Superintendent of Documents	Her Majesty's Stationery Office
U.S. Government Printing Office	P.O. Box 276
Washington, D.C. 20402, USA	London SW8 5DT, England

---

[213] This is usually referred to as the *RASC Handbook* since it is produced by the Royal Astronomical Society of Canada.

[214] The *Astronomical Almanac* is published jointly by the Royal Greenwich Observatory and the US Naval Observatory. Both addresses are given. It is also available as a computer disk.

Computer nets

Computers can talk to each other over telephone lines using an accessory called a modem. Often an individual or organization will set up a dedicated computer storing astronomical data or discussions. Astronomers may then dial in, extract the data and leave messages for other users. Many local or regional computer nets (more commonly called computer bulletin boards) are associated with local astronomy clubs or planetariums. The larger commercial nets, which charge an hourly fee for connection, usually have an astronomy special interest group, conference or topic where astronomers can chat and discuss their favorite subject. They also have large data bases with star catalogues, astronomy related software and calendar information on when to observe certain objects of interest. Some of the commercial nets also have digitized astronomical images from either their own users or from public sources such as NASA's Viking, Voyager, Einstein and Hubble spacecraft. At least one net (BIX) posts the IAU Circulars which announce the discovery of new comets, supernovae, flare stars, etc. Since almost all the commercial and international nets are cross-linked through Internet, any astronomer can reach any other astronomer electronically providing he knows the other's address. You can reach me as pete-manly@bix.com or as pmanly@mcimail.com. Note that all of the networks listed below are based in North America except CIX, which is in England.

America On-line; 703-893-6288, 800-827-6364 in Canada
BIX; 800-227-2983
CIX; 081 390 1255
CompuServe; 800-848-8199
GEnie; 800-638-9636

Organizations

Most of the organizations below publish journals at least quarterly containing recent developments, upcoming events and observing tips. One can become quite swamped with astronomical information by belonging to all of these organizations.

American Association of Variable Star Observers (AAVSO)
25 Birch St.
Cambridge, MA 02138, USA
617-354-0484, Fax: 617-354-0665

American Astronomical Society (AAS)
2000 Florida Avenue N.W., Suite 300
Washington, D.C. 20009, USA
202-328-2010

Association des Groupes d'Astronomes Amateurs
4545 Pierre-de-Coubertin, C. P. 1000
Succ. M
Montréal, Québec
Canada H7V 3R2

Association of Lunar and Planetary Observers (ALPO)
Harry D. Jamieson, Membership Secretary
P.O. Box 143
Heber Springs, AR 72543, USA
501-362-7264

Astronomical League
6235 Omie Circle
Pensacola, FL 32504, USA
904-477-8859

Astronomical Society of the Pacific (ASP)
390 Ashton Ave.
San Francisco, CA 94122, USA
415-337-1100, Fax: 415-337-5205, hotline recording 415-337-1244

British Astronomical Association (BAA)
Burlington House
Piccadilly, London W1V 0NL, England

International Amateur-Professional Photoelectric Photometry
(IAPPP)
Dyer Observatory, Vanderbilt University
Nashville, TN 37235, USA
615-373-4897

International Amateur-Professional Photoelectric Photometry (IAPPP)
Rolling Ridge Observatory
P.O. Box 8125
Piscataway, NJ 08854, USA
908-968-6025
CompuServe: 73760,303

International Astronomical Union (IAU)[215]
Central Bureau for Astronomical Telegrams (IAU Circulars)
Smithsonian Astrophysical Observatory
60 Garden St.
Cambridge, MA 02138, USA
TWX 719-320-6842 ASTROGRAM CAM
Telex # 921428
Internet: marsden@cfa.harvard.edu

International Dark Sky Association (IDSA)
950 N. Cherry Ave.
Tucson, AZ 85726, USA

International Occultation Timing Association (IOTA)
2760 SW Jewell Ave.
Topeka, KS 66611-1614, USA

National Deep Sky Observer's Society
1607 Washington Boulevard
Louisville, KY 40242, USA
502-426-4399

National Research Council
Institute of National Measurement Standards
Time and Frequency Standards Section
Building M36,
Montreal Road, Ottawa
Canada K1A 0R6
613-745-1576 (English), 613-745-9426 (French)

---

[215] The International Astronomical Union's Central Bureau for Astronomical Telegrams notifies astronomers world-wide via the IAU Circulars of new comet discoveries, supernovae, etc. For a yearly fee, observers may receive mailed Circulars or have them delivered electronically via many of the computer nets.

Royal Astronomical Society of Canada
136 Dupont Street,
Toronto, Ontario
Canada M5R 1V2
416-924-7973
Internet: asc@vela.astro.utoronto.ca

The Planetary Society
65 North Catalina Avenue
Pasadena, CA 91106, USA
818-793-5100

United States Department of Commerce
National Institutes of Standards and Technology
Radio Stations WWV/WWVB
2000 East County Road 58
Fort Collins, CO 80524, USA

Manufacturers

Astro Link
P.O. Box 1978
Spring Valley, CA 92077, USA
619-449-4722

Celestron International
P.O. Box 3578, 2835 Columbia St.
Torrance, CA 90503, USA
310-328-9560, Fax: 310-212-5835

Coyotè Enterprises
P.O. Box 22
Des Moines, NM 88418, USA

DayStar Filter Corporation
P.O. Box 1290
Pomona, CA 91769, USA
714-591-4673

Epoch Instruments
2331 American Ave.
Hayward, CA 94545, USA
510-784-0391

Lichtenknecker Optics N.V.
Kuringersteenweg 44
B-3500 Hasselt
Belgium

Lumicon Filters
2111Research Drive #5
Livermore, CA 94550, USA
800-767-9576, 510-447-9570, Fax: 510-447-9589

Meade Instruments Corporation
1675 Toronto Way
Costa Mesa, CA 92626, USA
714-556-2291, Fax: 714-556-4604

Optec, Inc.
199 Smith St.
Lowell, MI 49331, USA
616-897-9351, Fax: 616-897-8229

Orion Optics
Unit 12, Quakers Coppice
Crewe Gates Farm Industrial Estate
Crewe, Cheshire, CW1 1FA, England
0270 500089

Orion Telescope Center
2450 17th Ave., P.O. Box 1158
Santa Cruz, CA 95061-1158, USA
800-447-1001, 408-464-0446, Fax: 408-464-0466

Photometrics Ltd.
3440 E. Britannia Dr.
Tucson, AZ 85706, USA
602-889-9933
Fax: 602-573-1944

Santa Barbara Instrument Group
1482 East Valley Road #601
Santa Barbara, CA 93108, USA
805-691-1851, Fax: 805-969-0299

SpectraSource Instruments
31324 Via Colinas, Suite 114
Westlake Village, CA 91326, USA
818-707-2655, Fax: 818-707-9035

Texas Nautical Repair Co.
3110 South Shepherd
Houston, TX 77098, USA
713-529-3551, Fax: 713-529-3108, 800-880-3551

Telrad Incorporated
7092 Betty Dr.
Huntington Beach, CA 92647, USA
714-847-8903

Thousand Oaks Optical Filters
Box 5044-289
Thousand Oaks, CA 91359, USA
805-491-3642

Appendix 2
How to align the polar axis with the Earth's axis of rotation

The object of this exercise is to make the telescope's polar axis parallel to the Earth's rotation axis. Like a theological problem, there are many different paths to accomplish this. Like a theological problem, you will find people arguing incessantly as to the best method. The precision and detail of polar alignment has been taken to a fine art. There is probably a method described in your owner's manual. I will present another one here. You may wind up inventing your own.

Contrary to popular belief, the tripod does not have to be level, although this makes adjustments easier.[216] Note; severely leaning tripods can tip over.

The question arises as to how accurately you should align your telescope. Greater accuracy in alignment requires more set-up time and this is time which might have been spent observing. Clearly, spending an hour and a half aligning the telescope to observe the Moon visually at low power is excessive. On the other hand, if you're looking for Messier objects using the setting circles then a bad polar alignment is going to cost you search time every time you point the scope at blank sky and have to hunt around for a few minutes for each object. If you've just rolled the telescope out for a quick visual Lunar session while your spouse finishes dressing for an evening out, get within five degrees of the pole and start observing. If you're setting up for

---

[216] At one time I saw two astrophotographers argue this question to the point of a wager. A 20-cm S–C was set up with the top of the tripod fully 15° from level with the lowest point in the plane formed by the top of the tripod pointing to the Northwest. The telescope owner then proceeded to align the polar axis using only the azimuth and elevation adjustments on the telescope's wedge. He then made a fairly good one-hour astrophoto by the method of guiding via an off-axis eyepiece. The right ascension and declination corrections he needed were about the same as for a telescope with a leveled tripod.

astrophotography, spend 20 or 30 minutes before starting your one-hour exposures. If you are aligning a permanent pier prior to welding it down, I'd advise a minimum of three nights of work.

In aligning the telescope with the pole, I use the optical system of the telescope itself as a surveying instrument. First, I set the tripod down and point it approximately toward North, say within a couple of degrees. I do this by sighting along the side of the wedge at Polaris.[217] It's not very accurate but for the next few minutes of self-calibration of the optics, I'll be observing visually and it won't matter.

Then I swing the telescope around to a bright star and check the collimation of the optics. My telescope has fairly stiff collimation but some instruments require tweaking every time they're transported. The object of collimating the telescope is to ensure that the optical axis of the telescope (the path of the photons) is parallel to the mechanical tube of the telescope. If your telescope requires collimation, see the procedure in Appendix 3. After collimation, I also check the alignment of the finder telescope with respect to the main optics.[218] Finally, I check to make sure the crosshair reticle in the finder scope is rotated so that the hairs are parallel to right ascension and declination motion. This will come in handy later.

The next step is to swing the telescope back to where it is pointing roughly at the North Pole. The telescope tube should be roughly parallel to the fork arms. Set the right ascension so that the forks are on the East and West sides of the tube at about the same height. Using a star diagonal mirror, put a low-power wide-field eyepiece in the main optics and focus on whatever stars appear there. An eyepiece with a cross hair works best but

---

[217] For Southern Hemisphere observers the Mv 5 star Sigma Octantis can be used. The star has at times been called Polaris Australis.

[218] Some right-angle finder telescopes allow the diagonal mirror and eyepiece assembly to rotate with respect to the tube of the finder scope. While this makes for more convenient head angles when viewing, it can introduce a problem. Unless the rotating joint is exactly centered on and parallel with the finder scope's optical axis and the diagonal mirror's face is precisely 45° off the optical axis then rotating the diagonal and eyepiece will result in an apparent shift in the pointing of the cross hairs. Many inexpensive finder telescopes (including mine) have this problem. The cure is to set the diagonal once and not rotate it during the entire alignment procedure.

[219] For we folk who are short, this usually involves standing on something. I have a nice, sturdy military surplus ammunition case which also doubles as a carrying container for my tool kit, eyepieces and accessories.

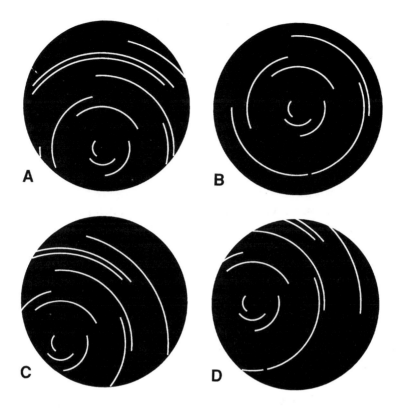

Figure A1. Polar alignment views through the finder telescope.

isn't absolutely necessary. Standing at the South side of the telescope (North side for Southern Hemisphere observers), lean over the telescope and look in the eyepiece.[219]

Lock the declination axis clamp so that the telescope does not move in that axis. After disengaging the right ascension clutch, rotate the telescope tube rapidly about its own axis. The stars in the eyepiece should appear to describe circles as shown in Fig. A1 A. Adjust the declination slow motion control until the stars describe circles which are concentric with the center of the field of view, as shown in Fig. A1 B. You can now be assured that the mechanical alignment of the polar axis is parallel to the optical axis.

If the stars appear to rotate as in Fig. A2 C then you have a problem. Either your collimation is way off or the right ascension and declination axes are not mutually perpendicular or the optical axis is not perpendicular to the declination axis. Nine times out of ten it is a problem of collimation. I'd advise going back and recollimating the telescope, as described in Appendix 3. If it is a problem of perpendicularity of axes then this is not something you can investigate in the dark.

The level-table method of sorting out this problem requires a level, stable surface and a dial indicator as found in many machine shops. The procedure is as follows; set telescope and base without tripod or wedge on level table (a surface plate found in machine shops is ideal). Set declination to nominally 90° North. Set feeler gauge on rim of front ring. Rotate in RA and adjust declination until you can get the gauge to indicate constant to within a few thousandths over the whole rotation. If you can, it's not a problem of the declination and RA axes not being perpendicular. If it is, some telescopes' fork arms can be removed and shimmed. This is rare but in a new telescope it is just cause for returning it to the manufacturer for realignment. Note; check roundness of the nose ring first. Do this by measuring the diameter of the ring with a caliper at several places. Now do a horizontal tube test; put a straight-through eyepiece in the telescope (still on flat plate) and focus it on a mark on the wall with the telescope tube roughly horizontal. Lock the RA clamp. Swing the telescope front aperture upward in declination and over the top until it's pointing at the opposite wall. Make a mark on the wall. Loosen the RA clamp and rotate in RA until the telescope is pointing once again at the first mark. Tighten the RA clamp. Loosen the declination clamp and swing the telescope front aperture upward in declination and over the top until it's pointing at the opposite wall. The telescope should be pointing at the mark to within a tenth of a degree, at worst. If it is not then the optical axis is not parallel with the mechanical axis of the telescope tube. Unless you suspect that you have a poor technique in collimating, then either one of the mirrors is not mounted concentrically on the mechanical axis of the tube, the primary mirror is tilted with respect to the tube axis or one of the mirrors was ground with an off-axis curve. Any of these problems in a new telescope are cause for returning the telescope to the manufacturer for repair under warranty. Note; before doing this ask yourself honestly if the telescope has been dropped since it was purchased. If so, then the manufacturer can usually repair it but he's going to want to be paid for his efforts.

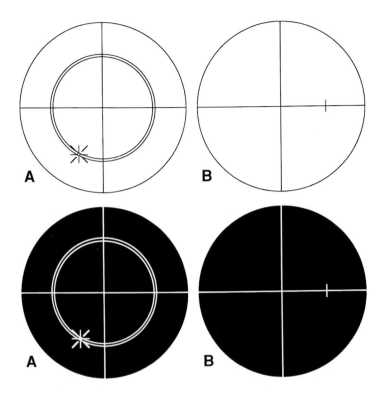

Figure A2. Polar alignmen t reticles for finder scopes.

Even if the stars describe circles as shown in Fig. A1 C, all is not lost. You can still align the telescope well enough for visual work. Adjust the declination until the stars describe arcs like those shown in Fig. A1 D.

With the optical axis of the telescope parallel to the right ascension mechanical axis, remember that the optical axis of the finder had been previously aligned to the main optics. Thus, the finder and the polar axis of the telescope are parallel. Without adjusting either the RA or declination axes, all that remains is to adjust the azimuth and elevation of the wedge until the finder scope points at Polaris or, for Southern Hemisphere observers, Sigma Octantis.

Adjustment in azimuth for some of the older telescopes meant kicking and shoving the tripod legs around until the wedge pointed North. More modern wedges have elongated arc-shaped slots for the azimuth bolts which make this much easier. One of these days I'm going to take a file to the three azimuth bolt holes on my old C-8 and get rid of the problem.

In actuality, one does not align the finder telescope precisely with the pole star since Polaris and Sigma Octantis are not exactly on the imaginary point which is an extension of the Earth's rotational axis. In fact, pointing of the Earth's axis moves with respect to the stars, describing a cone 46° across over a period of about 26,000 years. For the next century or so, however, Polaris will be fairly close to the pole. It is about 45 arc minutes from the true pole (Earth's axis) at a right ascension of 2h31'48.704' and a declination of +89°15'50.72' in Epoch 2000 coordinates. My finder scope has a reticle in it which has a ring whose radius is 45 arc minutes, as shown in Fig. A2 A.[220] Some finders simply have a tick mark on one or more of the cross hair legs which indicate a 45 arc minute angular distance, as shown in Fig. A2 B. For Southern hemisphere observers a similar reticle with a radius of about one degree can be made for Sigma Octantis which is located at 21h08'46.202' and –88°57'23.38'.

One must know not only the distance of the true pole from Polaris, but the direction with respect to the horizon. While the distance changes very slowly, the direction changes as the sky rotates with the time of night and the season. The chart in Fig. A3 shows the North circumpolar region. If you draw an imaginary line from Alkaid, the last star in the handle of the Big Dipper (Ursa Major) to the Eastern-most star in Cassiopeia (Epsilon Cassiopeiæ) then the line goes through the true pole and Polaris. The true pole is on the dipper side of Polaris.

I hold the chart in Fig. A3 up to the sky and rotate it until the line connecting Epsilon Cassiopeiæ, Polaris and Alkaid is parallel to the corresponding line in the sky. Then I note the angle that that line makes with the horizon. I find it is handy to measure hours in clock angle. For instance, in the example shown, the line appears to run from about the one o'clock position to the seven o'clock position.

I then set the right ascension on my telescope so that the lower fork arm beam is horizontal. Thus, the reticle cross hairs in my finder scope will describe a line parallel to the horizon and

---

[220] Well, it's actually 50 arc minutes but I made it 20 years ago when Polaris was 50 arc minutes from the pole and it's close enough.

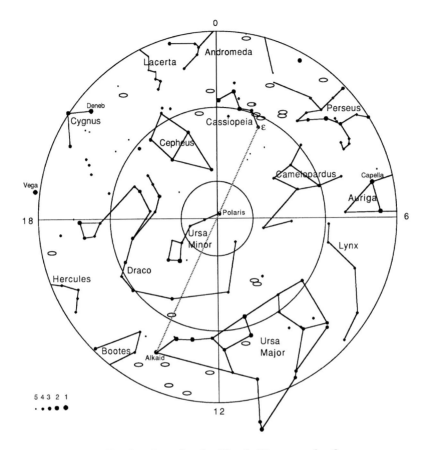

Figure A3. Finder chart for the North Circumpolar Sequence.

another vertical line. Then I adjust the altitude and azimuth of the wedge until Polaris falls on the circular ring and is in the one o'clock position. You need not estimate clock position closer than about a quarter of an hour. This would result in an hour angle error of 7.5° but this translates to an absolute error in misalignment of the telescope axis of about 0.09°.

At this point in polar alignment, visual observers may start looking at more interesting things. The telescope is aligned to better than 0.25° and setting circles may be used to find objects of interest. After a few nights of repeating this procedure, most observers should be able to get this far in three to five minutes. It should be noted that in the process, the telescope has been collimated and the finder has been aligned.

Some astrophotographers may be able to get along with an alignment error of 0.25° but guiding will probably involve considerable declination adjustment. The following procedure will generally produce an alignment of better than 0.1°.

First balance the telescope for a single eyepiece. Then put a high-power eyepiece in the main telescope tube. An eyepiece with cross hairs works best but it isn't absolutely necessary. I use the guiding eyepiece which has double cross hairs. Now focus on a star near the Celestial Equator on your local meridian. Spend two or three minutes adjusting your drive corrector so that its rate is approximately correct. Set the star exactly on the cross hairs and observe for five or ten minutes without adjusting the declination. If the star drifts toward the South the polar axis is pointing too far East.[221] Conversely, if the star drifts North then the polar axis is pointing too far West. Adjust it slightly. I find it handy to record on a piece of scrap paper how far the star moved South (or North) and how much adjustment of the tripod legs I made.[222] This will give you a feel for how much adjustment is required. For gross corrections, you may want to repeat the procedure.

Now focus on a star near the Eastern (or Western) horizon on the Celestial Equator. If the star drifts toward the South (or North if viewing on the Western horizon) then the polar axis is too low. Adjust the elevation of the polar axis. I generally repeat the entire procedure, going back to the meridian and then the horizon twice. It generally takes me an hour to complete alignment although I've seen it done in half that time with better precision, sometimes approaching 0.05° to 0.10°. Note that the procedure does not depend on the rate of the drive corrector, but only on observing the North/South drift of the star.

Before installing a telescope permanently and welding the pier in place, the same procedure is used but the star is allowed to drift for an hour or more before making adjustments. Stars should also be observed on both the East and West horizons in order to average out the effects of atmospheric refraction at

---

[221] For observers in the Southern Hemisphere, polar alignment by watching star drift is slightly different. In order to maintain the correct sense, swap the words 'North' and 'South' for the entire star drift procedure.

[222] A spread-sheet approach to this is described in *Sky & Telescope Magazine*, September, 1991, p. 299.

lower viewing elevations. This is one of those adjustments best suited to one who is terminally finicky. It requires great patience. On the other hand, when you are bored out of your mind waiting for one- or two-hour drift measurements-to-time out and you are tempted to say the alignment is good enough for a permanent mount, visualize yourself with a cutting torch going after the welds on the pier in order to make that one last adjustment you thought you didn't need earlier.

Appendix 3
Collimation of an S–C telescope

Note that more alignments have been ruined with sloppy collimation than have been fixed. Collimation of a good telescope should be approached with the same trepidation as changing the battery in your car. It's OK if you know how to do it but if you don't then it's best to have somebody show you how. The collimation screws in my telescope are fairly stiff and the instrument seldom requires adjustment but I have seen telescopes in which merely packing and unpacking the telescope from its carrying case is enough to throw the alignment off.

The object of collimation is to ensure that the three optical elements (corrector, primary and secondary) are all mutually centered on the optical axis and are parallel to each other. It is also important to ensure that these three elements are aligned properly with the baffle tube and mechanical axes of the telescope. For most commercially made Schmidt–Cassegrains, the primary mirror is not adjustable in tilt. Similarly, the corrector plate is firmly fixed to the front end of the tube and thus non-adjustable. The secondary mirror, however, is adjustable via three small screws mounted on the backside of the mirror support. In some telescopes the screws may have a plastic cap over them to keep out dust. If there is a fourth, central screw then do not touch that since it keeps the secondary from falling on the primary.

The first step in checking collimation requires fabrication or purchase of a collimating tool. The tool is made from a plastic 35-mm film canister which has had the bottom cut off and a 1–2-mm hole drilled in the cap, as shown in Fig. A4. The canister has a slight taper in its diameter which allows it to slip snugly into a standard 3.175-cm (1 1/4-inch) eyepiece holder. The tool ensures that when you look up the telescope tube, your eye is centered on the optical axis. Remove any eyepieces and star diagonal mirrors and insert the collimation tool directly into the

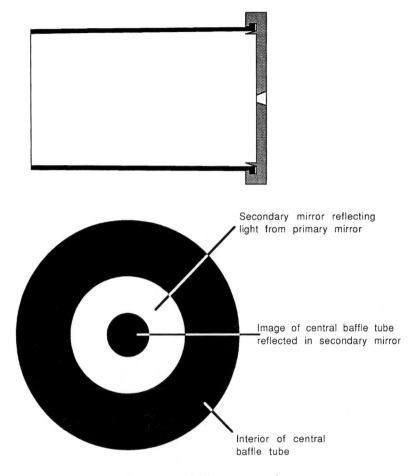

Secondary mirror reflecting
light from primary mirror

Image of central baffle tube
reflected in secondary mirror

Interior of central
baffle tube

Figure A4. Collimation tool.

rear of the telescope. Point the telescope at the daytime sky
(avoiding the Sun, of course) or a well-lit light-colored ceiling.
The view through the tool should be like that in Fig. A4. All of
the elements should appear concentric. If they are not, then a
preliminary adjustment of the secondary mirror may bring them
back into alignment.

Note: when collimating, make only small adjustments
(1/4 turn of adjusting screws or less at a time). If the
screw becomes hard to turn in a given direction, it may have
reached the end of its travel. Turn the other two screws in the
opposite direction to make the same adjustment. For most

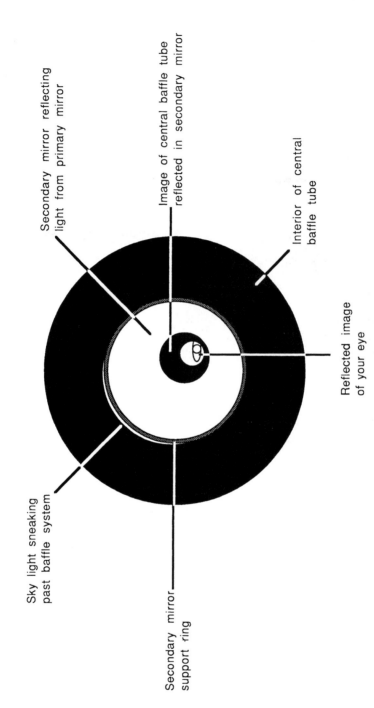

Secondary mirror reflecting light from primary mirror

Image of central baffle tube reflected in secondary mirror

Interior of central baffle tube

Reflected image of your eye

Sky light sneaking past baffle system

Secondary mirror support ring

Figure A5. View when collimating the telescope.

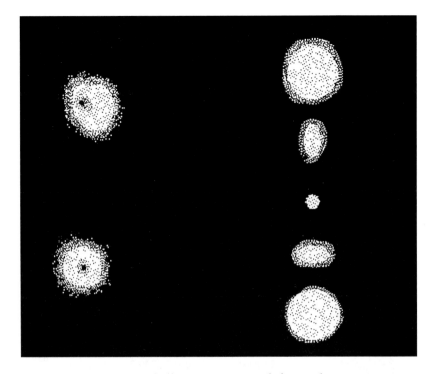

Figure A6. Collimation using a defocused star.

telescopes, once this preliminary daylight collimation is complete, it need not be repeated unless the telescope has been severely jostled.

Once the preliminary alignment is complete, you might want to check the baffling. This is a one-time inspection usually performed just after the scope's purchase. Remove the collimation tool and look in the open end of the baffle tube. Move your head from side to side so that you are looking from the side limit of the baffle tube. The view should look something like Fig. A5. The optical elements will normally appear non-concentric since your eye is off the optical axis. Look for any light sneaking past the secondary holder as shown in the illustration. Several models of 20-cm S–C telescope have this defect, which can be cured with a circle of black paper taped on the front of the secondary holder. If the amount of direct sky light going down the baffle tube is small then it may not be a problem except in Lunar and bright-planet viewing. If, however, you intend to do any daylight terrestrial or

Solar work then this little problem must be fixed or the contrast of images will suffer to an unacceptable point.

Having performed a preliminary collimation, the next step uses a bright star. Point the telescope and track the star using a medium-power eyepiece. While the recommended method is to place the eyepiece directly in the rear of the telescope, some folks use a star diagonal to make the viewing angle a bit more comfortable. Unfortunately, this also flips the image and makes the relationship between which collimation screw to turn and the image orientation a bit ambiguous. When the star is defocused it should appear as in the lower left corner of Fig. A6. This is the classical doughnut shape with a bright ring and a dark center. The dark portion should be concentric with the bright ring. If it is not, as shown in the upper left corner of Fig. A6, adjust the collimation screws until it is. Note: turning any of the screws will simultaneously move the dark area with respect to the brighter ring and move the entire image to one side. After each adjustment, therefore, the image of the star must be re-centered in the field of view.

On the first occasion of collimating the telescope or before purchasing the instrument, an astigmatism check should be performed. Using a relatively high-powered eyepiece, a series of stars of several different brightnesses should slowly be defocused and the image examined. As the focus knob is turned the image will get smaller, reaching some minimum. Continue turning the focus knob and examine the image as it expands in size. The images shown on the right side of Fig. A6 show a star which becomes elongated in one direction, then becomes a small fuzzy circle and then as optimum focus is passed, becomes an elongated image oriented 90° from the earlier view. Elongated images 90° from each other on either side of optimum focus indicate that one or more of the optical elements has an astigmatism. This can be indicative of a poorly made mirror or one of the optical elements, say the secondary, being stressed in its mounting. Noticeable astigmatism indicates a serious defect in the telescope, probably warranting return of the instrument to the dealer or factory for repair.

Appendix 4
Cleaning the corrector plate

Before cleaning the corrector plate, go back and re-read the last few paragraphs of Chapter 2. Then sit down and read a book or make a hot fudge sundae. Maybe the desire to clean the corrector plate will go away by the time you're finished. Cleaning has more potential for permanent damage to the telescope than any other operation. In the nearly 20 years that I have owned a 20-cm S–C I have had to clean the corrector only once and that was when I allowed heavy dew to form on the telescope. The dew contained industrial chemicals from a nearby city which created a gray-green film. I had also been warned that some of the chemicals could etch the glass. It would have been smarter of me to prevent the dew forming in the first place.

First blow off the dust using 'canned air' since you don't want any of these loose particles adhering to the glass when you wet it. Next, fill a spray bottle of the type used to clean windows with pure distilled water. With the corrector plate vertical (tube horizontal), spray the glass all over. The object here is to allow the water to dissolve all it can and carry off the dirt. Do not rub the glass since this can grind small dust particles into the glass and produce permanent scratches. Use plenty of water to allow it to flush away the grime.

At this point you may have completed the cleaning. Let the glass air-dry in a dust-free environment and then check it again. If deposits remain, repeat the flushing procedure with a mixture of one third isopropyl alcohol to two thirds distilled water. Note that the alcohol will leave a residue when it evaporates so after flushing with the alcohol/water mixture you must rinse with pure distilled water to get all of the alcohol off of the glass before it evaporates.

For particularly obstreperous grime, some people have added one drop of liquid dish detergent per liter of alcohol/water mixture. This mixture also leaves its own residue and must be rinsed off with pure water before the mixture evaporates.

For the toughest of particulate matter, mechanical action may be used sparingly. Remember that you risk scratching the corrector when you rub. Thus, dabbing at the corrector rather than wiping is recommended. Some people use wadded up non-silicone photographic lens cleaning tissues. Sterile cotton balls available from a drug store (chemist in England) are often inappropriate since some brands contain significant gritty (if sterile) particles. Clean cotton available at optical supply houses may not be sterile but it's free of grit. I use Q-tips®, small cotton swabs on the ends of soft wooden dowels.

After flushing the corrector plate thoroughly, leave the telescope tube horizontal and inspect the inside of the tube visually to ensure that water has not seeped between the corrector plate retaining ring and the glass, ending up inside the tube. If it has, leave the rear dust cap off the telescope and let the water evaporate. This may take several hours and should be done in a dust-free environment.

Recently commercial glass cleaners became available which consist of a viscous liquid which is poured onto the glass and allowed to dry into a rubbery mat. The mat is then pulled gently from the glass, carrying with it all of the grime and dust. While I have seen impressive demonstrations of these products I haven't tried them myself because they are fairly expensive, costing $50.00 – $100.00 per application.

Appendix 5
Mount vibrations

The S–C telescope is a mechanical system with moving parts. As such, it is subject to vibrations induced by observers bumping the telescope, cars and trucks rumbling by nearby and even the workings of the telescope motors and accessories such as camera shutters. As the telescope vibrates it shifts its pointing angle slightly, resulting in the eyepiece image moving back and forth. The first rule in vibration reduction is that you can never totally eliminate vibrations. You can minimize them to such a low level that they no longer bother you but they will always be there.

Given that a telescope is 'shaky' there are several cures but first you must determine where the vibration originates. If you can grab the telescope at the eyepiece end, shake it gently and you hear the bearings rattling then moving parts of your telescope are too loosely coupled together. First look for loose screws securing the tube to the declination bearing plates and the fork arms to the base. Some makes and model types have adjustable bearings which can be tightened. Care should be exercised in tightening bearings since excessive tightness will make the telescope hard to move, especially in cold weather. Furthermore, overly tight bearings are merely high-efficiency transmitters of higher-frequency vibrations.

Next, look at the mount. If a tripod is used, ensure that all of the adjustment screws are tight. On the older Celestron tripods, as shown in Fig. 5.16, there is an anti-vibration adjustment screw in the top center section of each leg. The screw should be adjusted so that its tip presses firmly against the tripod head. This puts a slight tension on the spreader bars at the bottom of the legs and thus any slop or play in the hinges of the legs will be taken up. There are similar devices on other manufacturers' tripods.

If you have a permanent pier you might want to examine resonances which can be produced in the pier. For example, I have seen an excellent steel pipe pier made from well-casing material sunk into a concrete pad. A short pipe section welded at the proper angle on the top provided a base for the telescope. It was sturdy, strong and when struck lightly with a hammer would ring like a bell for ten or fifteen seconds. The cure in this case was to fill the hollow pier with sand, a vibration dampening material.

With the telescope mechanically tight there are still some vibration-abatement tricks to be used. First, the nature of the vibration must be examined. All vibrations have a characteristic frequency. High-frequency vibrations result from quick, sharp motions like the opening of a camera shutter. Lower-frequency vibrations come from bumping the telescope or heavy equipment operating nearby. Determine if your problem is a high or low frequency one. Next, determine how quickly the vibrations dampen out when the source of vibrations is removed. For instance, look through the eyepiece and gently tap the front end of the telescope. A star's image will momentarily oscillate all over the field of view and then settle down. A half-second settling time is normal. If it persists for some time then you may have a vibration resonance with some element of your telescope. This is much like the ringing pier discussed above.

Persistent vibrations are usually, but not always, associated with higher frequencies. Resonances at these frequencies require stiff structures and tight mechanical coupling among vibrating members. One possible cure is to change the mechanical coupling between two resonating members such as the tube and fork arms. If you have tightened the bearings, try loosening the declination bearing just a little. This will change the coupling between the two vibrating pieces and may increase the dampening. Finally, if two members are resonating with respect to each other then you can change the resonant frequency of one and thus dampen the vibrations. For instance, Dan Ward has experimented with adding counterweights to the fork arms to change its resonant frequency. The fork is, after all, just a giant tuning fork and it will have its own natural vibration period. Judicious placement of small lead weights can, however, make it a very inefficient tuning fork.

If the source of your vibrations is external such as nearby traffic or vibrating machinery then you can decouple the entire telescope from its mount. An example is a nearby university with teaching telescopes situated on the roof of the science building to get them above the campus lights. Unfortunately, the building's air conditioning equipment is also on the roof and when it runs you can feel the vibrations through the soles of your shoes. The solution is to put either rubber pads like gymnastic mats or commercially available vibration isolators[223] under the legs of each tripod. Such measures are good only for low frequency vibrations in the 15–60 Hz region. Unfortunately, that allows the whole telescope and tripod assembly to vibrate at very low frequencies of 2–5 Hz when bumped. Thus, one problem is traded for another. Since the lower-frequency vibration dampens out after a few seconds the telescope is useable but the vibrations are annoying. When looked at as a trade-off, however, the dampened 2–5 Hz vibrations are much preferable to the constant 30 Hz vibrations from the air conditioner.

[223] One source of vibration isolation pads is Epoch Instruments. Their address is listed in Appendix 1.

Appendix 6
Field operations packing checklist

Your requirements will vary with your observing task and the local weather. For some of the more obscure items I have included a few words of explanation.[224]

Telescope optics assembly in packing case
Tripod
Wedge assembly
12 V 80 Amp-Hr battery
Drive corrector and associated cables
Dew shield
12 V d.c. defroster (for dew control)
Accessory box (cables, tools, telescope assembly bolts)
Eyepiece, filter and camera adapter box
Video kit and accessory box, spare tape (optional if making video observations)
Camera, accessory case, film (optional for photographic observations)
90 mm Maksutov spotting scope (for video or photographic observations)
Flashlight, spare flashlight, batteries, spare batteries
Telescope manual
Observing log
Norton's Star Atlas
Star charts, maps and finder charts for tonight's special objects
Pen, spare pen
Pocket programmable calculator
Adjustable-height observing chair

[224] I know of one observer whose checklist runs to two pages and includes an unpacking checklist to ensure that unused film is returned to the refrigerator to keep it fresh and data get reported as soon as possible.

Thermometer and wind gauge (occultation reports require this)
WWV radio receiver, spare batteries (for occultation timing)
Video time inserter (for occultations)
Solar filters (if camping for the weekend)
Jacket or sweater, wool observing hat (gloves optional)
Desert boots (rattlesnake resistant)
Spare glasses
Munchy food (check desert survival food and water already in car)
Camp stove, fuel (for making hot chocolate)
Folding lawn chair
Insect repellant and fly swatter
Bathroom tissue, paper towels and hand cleaner packets
Sleeping bag and camping gear (optional)
Astronomy club name badge
Check fuel in vehicle
Tell somebody where I'll be and when I'll return

Appendix 7
Astronomical nomenclature

The science of astronomy evolved from the practice of astrology and that, in turn, came from mythology and theology. Thus, we do not have a rigorous and logical set of names for the planets, stars and nebulae. We have a rag-tag collection of appellations derived from a score of cultures and epochs. At times during the last few centuries, astronomers have tried to rename everything to some sort of standardized system but this has met with only limited success, primarily because the newer naming systems are complex and many people still refer to the more popular objects with their older, more easily recognized names. Thus, the star Polaris (the North Star) is also known as Alpha Ursæ Minoris. This second name for the star reflects the system of designating stars by constellation and brightness. Polaris is the brightest star in the constellation Ursa Minor (Latin for the Little Bear). The second brightest star in that constellation is Beta Ursæ Minoris, the next brightest is Gamma Ursæ Minoris and so on.

This presents two immediate problems. First, the lettering is in a foreign language, Greek. Second, the spelling of the constellation is in yet another foreign language, Latin.[225] If you go looking for it on a star chart, you will not find anything labeled 'Ursæ Minoris', for that is the genitive Latin version of 'Ursa Minor', which is used on star charts. Just to make things interesting, nobody refers to that constellation as a bear, but rather as the Little Dipper. It looks like niether a bear nor a dipper to me. A kite, certainly, but a bear? Really! And for a final point of confusion, in some constellations like Ursa Major the stars are lettered

---

[225] I've always suspected collusion between the Astronomy Department and the Classics Department in most universities. They teach subjects in such a way that you always have to take classes in the other department just to decode what's being taught in the original class.

in sequence along the mythical figure and not by brightness.[226] We can clear up confusion about the Greek alphabet easily as follows

Alpha	α	Eta	η	Nu	ν	Tau	τ
Beta	β	Theta	θ	Xi	ξ	Upsilon	υ
Gamma	γ	Iota	ι	Omicron	o	Phi	φ
Delta	δ	Kappa	κ	Pi	π	Chi	χ
Epsilon	ε	Lambda	λ	Rho	ρ	Psi	ψ
Zeta	ζ	Mu	μ	Sigma	σ	Omega	ω

Generally, when writing about stars in text, the English word such as Alpha is used. On star charts the Greek letter, α in this case, is used. Just when you might think you have the system firmly in hand, along comes a new wrinkle. The stars are generally named by brightness using lower-case Greek letters but when astronomers found more stars than Greek letters they started to use lower-case Roman letters like a, b, c, etc. After those were exhausted they used upper-case Roman letters. For letters like Omicron and Chi, which resemble Roman letters, this becomes very confusing. The capitals after Q are not used but designations such as R, S and T were appropriated by the folks interested in variable stars to name things of interest only to them. After the alphabets, stars were simply numbered by increasing right ascension in each constellation. Such designations are called Flamsteed Numbers, after the astronomer who performed the prodigious task of numbering them all. Thus, we have the star 47 Tucanæ which is the 47th star in the constellation Tucanis.[227]

In modern times, computers have enabled us to name and classify many more stars than before. Thus, we have the Smithsonian Astrophysical Observatory Catalogue and Atlas of over a quarter million stars, each with its own separate number. I have held in my hand an optical computer disk of star positions with tens of millions of entries. While this may seem like more information than you would ever want to know, I have had to go to that data base on two occasions to find out about some star I could see in my 20-cm S–C telescope.

---

[226] *Norton's Star Atlas*, 18th ed, Arthur P. Norton and J. Inglis, edited by I. Ridpath, Sky Publishing Corporation, 1989 ISBN 0 85248 900 5, p. 88.

[227] We have a little problem with 47 Tucanæ in that it looks like a fifth-magnitude star only to the naked eye. In a telescope it appears as a magnificent globular cluster.

Just to make things one step more confusing, over the centuries some stars have brightened and some have faded and thus the Alpha of some constellation may be much fainter than the Beta. In this book I will use the most popular name I know such as Polaris. If there is any uncertainty I will add a second name such as Alpha Ursæ Minoris or more (North Star) if necessary. I do not expect the reader to know that this star is also called SAO 0003008 or BD +84° 0008.[228]

In nebulae and galaxies there is similar confusion. Herschel numbered and catalogued everything which passed before his eyepiece and thus some objects have Herschel Numbers. Messier made a list of things he wasn't interested in (non-comets) and thus we have the Messier Catalogue but some of these are gas clouds internal to our own galaxy while others are external galaxies in their own right. Messier's objects are preceded by an M with a number and are used heavily in this book. Then, for fainter objects the New General Catalogue (NGC) lists thousands of nebulous objects. Finally, many of the better viewing objects have acquired popular names like the Ring Nebula (M57). As with stellar nomenclature, I will give the most popular name and an alternate or two such as M43 (Orion Nebula).

The nomenclature of astronomical objects resembles murky water at times – the objects are there but you have to feel around for them sometimes. Elitists will occasionally use obscure terminology in order to appear erudite. As an example, I once had to spend several minutes looking up a reference in a little-known catalogue, only to find that the star in question was Betelgeuse, a well-known nearby red giant star. Another trick to watch for is the alternate or foreign spelling such as Betelgeux or Betelgeuze. I have had to work in several scientific fields from aeronautics to zoology and frankly, the taxonomy (classification and naming) in astronomy is the most hodgepodge, mixed-up, ill-conceived, patchwork conglomeration of miscellaneous jumbled odds and ends using a pot pourri of nouns and numbers that I have ever seen. But it could be worse, for I have never attempted a serious foray into politics.

---

[228] These two numbers are from the Smithsonian Astrophysical Observatory (SAO) Catalogue and the Bonner Durchmusterung des Nordlichen Himmels.

Appendix 8

Catalogue of bright stars and interesting things

RA	Dec	Proper Motion (RA)	(Dec)	Radial Vel.	Spect. Class	Dist. ly	Mv	Const.		Common name	Comments
00h08'23.265'	+29°05'25.58'	+1.039	-16.33	-12.0	B9p	127	2.06	α	And	Alpheratz	spectroscopic binary
00h26'17.030'	-42°18'21.81'	+1.833	-39.57	+75.0	K0 III	91	2.39	α	Phe	Ankaa	spectroscopic binary
00h40'30.450'	+56°32'14.46'	+0.636	-03.19	-04.0	K0 II-III	147	2.23	α	Cas	Schedar	
00h43'35.372'	-17°59'11.82'	+1.637	+03.25	+13.0	K1 III	59	2.04	β	Cet	Diphda	variable
01h37'42.852'	-57°14'12.18'	+1.173	-03.47	+16.0	B5 IV-V	127	0.46	α	Eri	Achernar	
02h07'10.403'	+23°27'44.66'	+1.383	-14.83	-14.0	K2 III	75	2.00	α	Ari	Hamal	variable
02h31'48.704'	+89°15'50.72'	+19.877	-01.52	-17.0	F8 Ib	782	2.02	α	UMi	Polaris	variable spectroscopic binary
02h44'11.986'	+49°13'42.48'	+3.425	-08.95	+25.0			4.12	θ	Per		
02h58'15.696'	-40°18'16.97'	-0.391	+01.94	+12.0			3.42	θ-1	Eri	Acamar	
03h02'16.773'	+04°05'22.93'	-0.063	-07.80	-26.0	M2 III	147	2.53	α	Cet	Menkar	
03h08'10.114'	+40°57'20.76'	+0.0003	+0.002	+04.0	B8 V	104	2.20	β	Per	Algol	eclipsing binary
03h24'19.365'	+49°51'40.34'	+0.246	-02.46	-02.0	F5 Ib	522	1.80	α	Per	Mirfak	variable
04h35'55.237'	+16°30'33.39'	+0.439	-18.97	+54.0	K5 III	68	0.85	α	Tau	Aldebaran	variable

RA	Dec	Proper Motion (RA)	(Dec)	Radial Vel.	Spect. Class	Dist. ly	Mv	Const.		Common name	Comments
05h14'32.268'	-08°12'05.98'	+0.003	-00.13	+21.0	B8 Ia	815	0.12	β	Ori	Rigel	variable
05h16'41.353'	+45°59'52.90'	+0.728	-42.47	+30.0	G8+F	46	0.08	α	Aur	Capella	spectroscopi binary
05h25'07.857'	+06°20'58.74'	-0.059	-01.39	+18.0	B2 III	303	1.64	γ	Ori	Bellatrix	variable
05h26'17.511'	+28°36'26.67'	+0.169	-17.51	+09.0	B7 III	179	1.65	β	Tau	Elnath	
05h36'12.809'	-01°12'07.02'	+0.006	-00.24	+26.0	B0 Ia	1532	1.70	ε	Ori	Alnilam	
05h55'10.307'	+07°24'25.35'	+0.173	+00.87	+21.0	M2 I	652	0.50	α	Ori	Betelgeuse	variable spectro-scopic binary
06h23'57.119'	-52°41'44.50'	+0.245	+02.07	+21.0	F0 Ib	196	-0.72	α	Car	Canopus	
06h45'08.871'	-16°42'57.99'	-3.847	-120.53	-07.6	A1 V	8.8	-1.46	α	CMa	Sirius	variable, double split
06h58'37.548'	-28°58'19.50'	+0.031	+00.28	+27.0	B2 II	652	1.50	ε	CMa	Adhara	
07h39'18.113'	+05°13'30.06'	-4.755	-102.29	-03.0	F5 IV	11	0.38	α	CMi	Procyon	variable
07h45'18.946'	+28°01'34.26'	-4.740	-04.59	+03.0	Ko III	36	1.14	β	Gem	Pollux	variable, nearest giant star
08h22'30.833'	-59°30'34.51'	-0.346	+01.44	+02.0	K0 II+B	326	1.86	ε	Car	Avior	variable
09h07'59.776'	-43°25'57.38'	-0.172	+01.27	+18.0	K5 Ib	85	2.21	λ	Vel	Suhail	variable
09h13'11.957'	-69°43'01.95'	-3.108	+10.78	-05.0	A0 III	600	1.68	β	Car	Miaplacidus	
09h27'35.247'	-08°39'31.15'	-0.093	+03.28	-04.0	K4 III	98	1.98	α	Hya	Alphard	variable
10h08'22.315'	+11°58'01.89'	-1.693	+00.64	+06.0	B7 V	85	1.35	α	Leo	Regulus	variable
11h03'43.666'	+61°45'03.22'	-1.675	-06.65	-09.0	K0 III	104	1.79	α	UMa	Dubhe	variable
11h49'03.580'	+14°34'19.35'	-3.422	-11.41	00.0	A3 V	42	2.14	β	Leo	Denebola	variable
12h15'48.366'	-17°32'30.97'	-1.124	+02.33	-04.0	B8 III	277	2.59	γ	Crv	Gienah	variable
12h26'35.871'	-63°05'56.58'	-0.524	-01.21	-11.0	B2 IV	261	1.58	α-1	Cru	Acrux	spectroscopic binary

RA	Dec				Spectral	Dist	Mag		Const	Name	Notes
12h29'06.8'	+02°03'07'	0	0	4.7E4		2.6E9	12.0		Vir	3C273	nearest & brightest quasar
12h31'09.929'	-57°06'47.50'	+0.285	-26.23	+21.0	M3 II	228	1.63	γ	Cru	Gacrux	variable
12h54'01.748'	+55°57'35.47'	+1.328	-00.58	-09.0	A0p	82	1.77	ε	UMa	Alioth	variable spectroscopic binary
13h25'11.587'	-11°09'40.71'	-0.278	-02.83	+01.0	B1 V	261	0.97	α	Vir	Spica	eclipsing spectroscopic binary
13h47'32.434'	+49°18'47.95'	-1.249	-01.09	-11.0	B3 V	147	1.86	η	UMa	Alkaid	variable
14h03'49.408'	-60°22'22.79'	-0.426	-01.93	+06.0	B1 II	391	0.61	β	Cen	Hadar	
14h06'40.951'	-36°22'12.03'	-4.293	-51.90	+01.0	K0 IV	55	2.06	θ	Cen	Menkent	
14h15'39.677'	+19°10'56.71'	-7.714	-199.84	-05.0	K2p III	36	-0.04	α	Boo	Arcturus	variable
14h39'36.087'	-60°50'07.14'	-49.486	+69.60	-22.2	M5	4.3			Cen	Proxima Centauri	(nearest star)
14h33'35.885'	-60°50'07.44'	-49.826	+69.93	-22.2	G2 V	4.4	-0.01	α	Cen	Rigil Kent	double star
14h50'52.713'	-16°02'30.42'	-0.734	-06.68	-10.0			2.75	α-2	Lib	Zubenelgen	
14h50'42.346'	+74°09'19.78'	-0.763	+01.22	+17.0	K4 III	104	2.08	β	UMi	Kochab	variable
15h34'41.276'	+26°42'52.94'	+0.906	-08.86	+02.0	A0 V	75	2.23	α	CrB	Alphecca	variable spectroscopic binary
16h29'24.439'	-26°25'55.15'	-0.071	-02.03	-03.0	M1 Ib	424	0.96	α	Sco	Antares	variable
16h48'39.869'	-69°01'39.82'	+0.260	-03.40	-03.0	K4 III	92	1.92	α	TrA	Atria	
17h10'22.681'	-15°43'29.71'	+0.260	+09.50	-01.0	A2 V	69	2.43	η	Oph	Sabik	
17h33'36.534'	-37°06'13.72'	-0.011	-02.92	-03.0	B1 V	326	1.63	λ	Sco	Shaula	spectroscopic binary
17h34'56.076'	+12°33'36.14'	+0.822	-22.64	+13.0	A5 III	59	2.08	α	Oph	Ras-Alhague	
17h56'36.367'	+51°29'20.21'	-0.081	-01.94	-28.0	K5 III	117	2.23	γ	Dra	Eltanin	variable
17h57'48'	+04°04'02'	-4.785	+1026	+141	dM5 V	5.9	9.54			Barnard's Star	(highest proper motion)

RA	Dec	Proper Motion (RA)	(Dec)	Radial Vel.	Spect. Class	Dist. ly	Mv	Const.		Common name	Comments
18h24'10.327'	-34°23'04.73'	-0.309	-12.41	-15.0	B9 IV	163	1.85	ε	Sgr	Kaus_Aus.	
18h36'56.332'	+38°47'01.17'	+1.726	+28.61	-14.0	A0 V	26	0.03	α	Lyr	Vega	variable
18h55'15.924'	-26°17'48.23'	+0.099	-05.42	-11.0	B2 V	261	2.02	σ	Sgr	Nunki	
19h30'43.241'	+27°57'34.72'	+0.0001	-0.002		B9			β	Cyg	Albireo	gold/blue-green double
19h50'47.002'	+08°52'06.03'	+3.629	+38.63	-26.3	A7 V	16	0.77	α	Aql	Altair	
20h25'38.852'	-56°44'06.38'	+0.082	-08.91	+02.0	B3 IV	293	1.94	α	Pav	Peacock	variable spectroscopic binary
20h41'25.917'	+45°16'49.31'	+0.027	+00.23	-05.0	A2 Ia	1630	1.25	α	Cyg	Deneb	variable
21h08'46.202'	-88°57'23.38'	+8.490	+00.47	+12.0			5.47	σ	Oct	Polaris Aus.	variable
21h44'11.164'	+09°52'29.92'	+0.207	-00.06	+05.0	K2 Ib	815	2.39	ε	Peg	Enif	variable
22h08'14.000'	-46°57'39.59'	+1.259	-15.10	+12.0	B5 V	68	1.74	α	Gru	Al_na'ir	variable
22h57'39.055'	-29°37'20.10'	+2.551	-16.47	+07.0	A3 V	23	1.16	α	PsA	Fomalhaut	
23h04'45.658'	+15°12'18.90'	+0.436	-04.25	-04.0	B9.5 III	108	2.49	a	Peg	Markab	variable

Epoch 2000 Bright Star or Otherwise Interesting Object Star Catalogue.

Sources; *Norton's Star Atlas, Allen's Astrophysical Quantities, SAO Catalogue, Astronomical Almanac*

Appendix 9
Catalogue of nearby stars

Star name	Dist Ly	RA hr	m	Dec Deg m	Mv	Sp	Mass Ms	Comments
Proxima Centauri	4.27	14	26	-62 28	15.45	M5	0.1	Bound to Alpha Centauri A & B, 9500 AU away
Alpha Centauri A	4.38	14	36	-60 38	04.35	G2	1.1	23 AU from B component, bound to Proxima
Alpha Centauri B	4.38	14	36	-60 38	05.69	K5	0.89	23 AU from A component, bound to Proxima
Barnard's Star	5.91	17	55	+04 33	13.25	M5		High proper motion
Wolf 359	7.60	10	54	+07 19	16.68	M8		Flare star
BD +36* 2147	8.13	11	01	+36 18	10.49	M2	0.35	Pair of dwarf double stars
Luyten 726-8	8.88	01	36	-18 13	15.27	M5	0.044	Double with UV Ceti
UV Ceti	8.88	01	36	-18 13	15.8	M6	0.035	Flare star
Sirius A	8.65	06	43	-16 39	-1.42	A1	2.31	Brightest star in our sky
Sirius B	8.65	06	43	-16 39	11.56	A	0.98	Dwarf companion to Sirius A
Ross 154	9.45	18	47	-23 53	13.3	M4		
Ross 248	10.3	23	39	+43 55	14.80	M6		
Ross 128	10.8	11	45	+01 06	13.50	M5		
Epsilon Eridani	10.8	03	31	-09 38	6.13	K2		Flare star

Star name	Dist Ly	RA hr	m	Dec Deg m	Mv	Sp	Mass Ms	Comments
Luyten 789-6	10.8	22	36	-15 36	14.6	M7		
61 Cygni A	11.1	21	05	+38 30	7.58	K5	0.63	Double with 61 Cygni B
61 Cygni B	11.1	21	05	+38 30	8.39	K7	0.6	Double with 61 Cygni A
Epsilon Indi	11.2	22	00	-57 00	7.00	K5		
Procyon A	11.4	07	37	+05 21	2.64	F5	1.77	Double with Procyon B
Procyon B	11.4	07	37	+05 21	13.0	F	0.63	Dwarf companion to Procyon A
Tau Ceti	11.8	01	41	-16 12	5.72	G8		
Wolf 28	13.8	00	46	+05 09	14.26	G	Dwarf	
Sigma Pav	18.6	20	04	-66 19	4.76	G6		
Eta Cas A	19.2	00	46	+57 33	4.60	G0	0.85	Separated by 70 AU from B component
Eta Cas B	19.2	00	46	+57 33	8.66	M0	0.52	Separated by 70 AU from A component
82 Eridani	20.2	03	17	-43 16	5.29	G5		
Beta Hyi	20.5	00	23	-77 32	3.80	G1	Dwarf	

The table lists all stars out to 12 light years and only types F6-G8 beyond that.

Appendix 10
Messier Catalogue

	RA	Dec	RV km/s	Dist. ly	Mv	Size	Type	Const.	Alternate names
M1	05h34.5'	+22°01'		7100	8.4	6'×4'	supernova remnant	Tau	NGC 1952, Crab Nebula, (pulsar)
M2	21h33.5'	-00°49'			6.5	12'	globular cluster	Aqr	NGC 7089
M3	13h42.2'	+28°23'		4.2E4	6.4	19'	globular cluster	CVn	NGC 5272
M4	16h23.6'	-26°32'		9130	5.9	20'	globular cluster	Sco	NGC 6121
M5	15h18.6'	+02°05		2.8E4	5.8	20'	globular cluster	Ser	NGC 5904
M6	17h40.1'	-32°12'		1956	4.6	26'	open cluster	Sco	NGC 6405, C1736-321
M7	17h54.0'	-34°48'		782	3.3	50'	open cluster	Sco	NGC 6475, C1750-348
M8	18h03.8'	-24°23'		3900	5.8	90'×40'	diffuse nebula	Sgr	NGC 6523, Lagoon Nebula
M9	17h19.2'	-18°31'			7.9	6'	globular cluster	Oph	NGC 6333
M10	16h57.1'	-04°06'			6.6	12'	globular cluster	Oph	NGC 6254
M11	18h51.1'	-06°16'		5575	6.1	12'	open cluster	Sct	NGC 6705, C1848-063, Wild Duck Nebula
M12	16h47.2'	-01°57'		1.6E4	6.6	12'	globular cluster	Oph	NGC 6218
M13	16h41.7'	+36°28'		2.5E4	5.9	23'	globular cluster	Her	NGC 6205, Hercules Cluster
M14	17h37.6'	-03°15'			7.6	7'	globular cluster	Oph	NGC 6402
M15	21h30.0'	+12°10'		4.6E4	6.4	12'	globular cluster	Peg	NGC 7078
M16	18h18.9	-13°47'		6850	6.5	8'	open cluster	Ser	NGC 6611, C1816-138

	RA	Dec	RV km/s	Dist. ly	Mv	Size	Type	Const.	Alternate names
M17	18h20.8'	-16°11'		5200	7.0	46' × 37'	diffuse nebula	Sgr	NGC 6618, Omega, Horseshoe or, Swan Nebula
M18	18h19.9'	-17°08'			6.9	7'	open cluster	Sgr	NGC 6613
M19	17h02.6'	-26°16'		2.3E4	7.2	5'	globular cluster	Oph	NGC 6273
M20	18h02.6'	-23°02'		3260	8.5	29' × 27'	diffuse nebula	Sgr	NGC 6514, Trifid Nebula
M21	18h04.6'	-22°30'		4075	7.2	12'	open cluster	Sgr	NGC 6531, C1801-225
M22	18h36.4'	-23°54'		9780	5.1	17'	globular cluster	Sgr	NGC 6656
M23	17h56.8'	-19°01'		2151	5.9	27'	open cluster	Sgr	NGC 6494, C1753-190
M24	18h16.9'	-18°29'			4.5	90'	star cloud	Sgr	
M25	18h31.7'	-19°14'		1890	6.2	35'	open cluster	Sgr	C1828-192, IC4725
M26	18h45.2'	-09°24'		5053	9.0	9'	open cluster	Sct	NGC 6694, C1842-094
M27	19h59.6'	+22°43'		717	8.1	8' × 4'	planetary nebula	Vul	NGC 6853, Dumbell Nebula
M28	18h24.5'	-24°52'			6.9	15'	globular cluster	Sgr	NGC 6626
M29	20h23.9'	+38°32'		4075	7.5	7'	open cluster	Cyg	NGC 6913, C2022+383
M30	21h40.4'	-23°11'			7.5	9'	globular cluster	Cap	NGC 7099
M31	00h42.7'	+41°16'	-229	2.2E6	4.4	160' × 40'	spiral galaxy	And	NGC 0224, Andromeda Galaxy or nebula
M32	00h42.7'	+40°52'	-217	2.2E6	9.1	3' × 2'	elliptical galaxy	And	NGC 0221
M33	01h33.9'	+30°39'	-183	2.4E6	6.3	60' × 40'	spiral galaxy	Tri	NGC 0598
M34	02h42.1'	+42°46'		1434	5.8	30'	open cluster	Per	NGC 1039, C0238+425
M35	06h08.8'	+24°21'		2836	5.6	29'	open cluster	Gem	NGC 2168, C0605+243
M36	05h36.1'	+34°07'		4100	6.5	16'	open cluster	Aur	NGC 1960, C0532+341
M37	05h52.5'	+32°33'		3900	6.2	24'	open cluster	Aur	NGC 2099, C0549+325
M38	05h27.7'	+35°51'		3900	6.8	18'	open cluster	Aur	NGC 1912, C0525+358
M39	21h32.2'	+48°27'		830	5.3	32'	open cluster	Cyg	NGC 7092, C2130+482

	RA	Dec		mag	size	type		
M40	12h22.4'	+58°05'		8.0		double star		
M41	06h47.1'	-20°44'	2282	5.0	32'	open cluster	CMa	NGC 2287, C0644-206
M42	05h35.4'	-05°27'	1500	4.0	66' × 60'	diffuse nebula	Ori	NGC 1976, Great Nebula in Orion
M43	05h35.6'	-05°16'		9.0		diffuse nebula	Ori	NGC 1982, NE wing of Orion Nebula
M44	08h40.1'	+19°59'	522	3.9	90'	open cluster	Cnc	NGC 2632, C0837+201, Praesepe or Beehive
M45	03h47.0'	+24°07'	410	1.5	120'	open cluster	Tau	C0344+239, Pleiades
M46	07h41.8'	-14°49'	5412	6.6	27'	open cluster	Pup	NGC 2437, C0739-147
M47	07h36.6'	-14°30'	1565	4.3	25'	open cluster	Pup	NGC 2422, C0734-143
M48	08h13.7'	-05°48'	1989	5.5	30'	open cluster	Hya	NGC 2548, C0811-056
M49	12h29.8'	+08°00'		8.4	4' × 4'	elliptical galaxy	Vir	NGC 4472
M50	07h03.2'	-08°20'	2967	7.2	16'	open cluster	Mon	NGC 2323, C0700-082
M51	13h29.9'	+47°12'	1.2E7	8.1	12' × 6'	spiral galaxy	CVn	NGC 5194-5, Whirlpool Galaxy
M52	23h24.2'	+61°35'	5216	8.2	13'	open cluster	Cas	NGC 7654, C2322+613
M53	13h12.9'	+18°10'	6.8E4	7.7	14'	globular cluster	Com	NGC 5024
M54	18h55.1'	-30°29'		7.7	6'	globular cluster	Sgr	NGC 6715
M55	19h40.0'	-30°58'	2.0E4	7.0	15'	globular cluster	Sgr	NGC 6809
M56	19h16.6'	+30°11'		8.2	5'	globular cluster	Lyr	NGC 6779
M57	18h53.6'	+33°02'	2280	9.0	1' × 1'	planetary nebula	Lyr	NGC 6720, Ring Nebula
M58	12h37.7'	+11°49'		9.8	4' × 3'	spiral galaxy	Vir	NGC 4579
M59	12h42.0'	+11°39'		9.8	3' × 2'	elliptical galaxy	Vir	NGC 4621
M60	12h43.7'	+11°33'		8.8	4' × 3'	elliptical galaxy	Vir	NGC 4649
M61	12h21.9'	+04°28'		9.7	6' × 6'	spiral galaxy	Vir	NGC 4303
M62	17h01.2'	-30°07'	2.6E4	6.6	6'	globular cluster	Oph	NGC 6266
M63	13h15.8'	+42°02'	1.5E7	8.6	8' × 3'	spiral galaxy	CVn	NGC 5055
M64	12h56.7'	+21°41'	1.3E7	8.5	8' × 4'	spiral galaxy	Com	NGC 4826, Blackeye Galaxy
M65	11h18.9'	+13°05'		9.3	8' × 2'	spiral galaxy	Leo	NGC 3623

	RA	Dec	RV km/s	Dist. ly	Mv	Size	Type	Const.	Alternate names
M66	11h20.2'	+12°59'	790		9.0	8' × 2'	spiral galaxy	Leo	NGC 3627
M67	08h51.2'	+11°49'		2700	7.4	18'	open cluster	Cnc	NGC 2682, C0847+120
M68	12h39.5'	-26°45'		3.8E4	8.2	9'	globular cluster	Hya	NGC 4590
M69	18h31.4'	-32°21'			7.7	4'	globular cluster	Sgr	NGC 6637
M70	18h43.2'	-32°18'			8.1	4'	globular cluster	Sgr	NGC 6681
M71	19h53.8'	+18°47'		2.7E4	8.3	6'	globular cluster	Sge	NGC 6838
M72	20h53.5'	-12°32'			9.4	5'	globular cluster	Aqr	NGC 6981
M73	20h58.9'	-12°38'					globular cluster	Aqr	NGC 6994
M74	01h36.7'	+15°47'			9.2	8' × 8'	spiral galaxy	Psc	NGC 628
M75	20h06.1'	-21°55'			8.6	5'	globular cluster	Sgr	NGC 6864
M76	01h42.4'	+51°34'			11.5	2' × 1'	planetary nebula	Per	NGC 650-1
M77	02h42.7'	-00°01'		5.2E7	8.8	2' × 2'	spiral galaxy	Cet	NGC 1068, (Seyfert galaxy)
M78	05h46.7'	+00°03'		2.0E5	8.0	8' × 6'	diffuse nebula	Ori	NGC 2068
M79	05h24.5'	-24°33'			8.0	8'	globular cluster	Lep	NGC 1904
M80	16h17.0'	-22°59'			7.2	5'	globular cluster	Sco	NGC 6093
M81	09h55.6'	+69°04'		1.0E7	6.8	16' × 10'	spiral galaxy	UMa	NGC 3031
M82	09h55.8'	+69°41'		9.9E6	8.4	7' × 2'	Irregular galaxy	UMa	NGC 3034, (radio galaxy)
M83	13h37.0'	-29°52'		1.0E7	7.6	10' × 8'	spiral galaxy	Hya	NGC 5236
M84	12h25.1'	+12°53'			9.3	3' × 3'	elliptical galaxy	Vir	NGC 4374
M85	12h25.4'	+18°11'			9.2	4' × 2'	elliptical galaxy	Com	NGC 4382
M86	12h26.2'	+12°57'			9.2	4' × 3'	elliptical galaxy	Vir	NGC 4406
M87	12h30.8'	+12°24'		4.2E7	8.6	3' × 3'	elliptical galaxy	Vir	NGC 4486, Radio Galaxy
M88	12h32.0'	+14°25'			9.5	6' × 3'	spiral galaxy	Com	NGC 4501

M	RA	Dec			Mag	Size	Type	Const	Designation
M89	12h35.7'	+12°33'			9.8	2' × 2'	elliptical galaxy	Vir	NGC 4552
M90	12h36.8'	+13°10'			9.5	6' × 3'	spiral galaxy	Vir	NGC 4569
M91	12h35.4'	+14°30'			10.2				NGC 4548
M92	17h17.1'	+43°08'		3.3E4	6.5	12'	globular cluster	Her	NGC 6341
M93	07h44.6'	-23°52'		3585	6.5	18'	open cluster	Pup	NGC 2447, C0742-237
M94	12h50.9'	+41°07'		1.5E7	8.1	5' × 4'	spiral galaxy	CVn	NGC 4736
M95	10h44.0'	+11°42'			9.7	3' × 3'	spiral galaxy	Leo	NGC 3351
M96	10h46.8'	+11°49'	790		9.2	7' × 4'	spiral galaxy	Leo	NGC 3368
M97	11h14.8'	+55°01'		1956	11.2	3' × 3'	planetary nebula	UMa	NGC 3587, Owl Nebula
M98	12h13.8'	+14°54'			10.1	8' × 2'	spiral galaxy	Cpm	NGC 4192
M99	12h18.8'	+14°25'			9.8	4' × 4'	spiral galaxy	Com	NGC 4254
M100	12h22.9'	+15°49'			9.4	5' × 5'	spiral galaxy	Com	NGC 4321
M101	14h03.2'	+54°21'	241	1.2E7	8.2	22' × 22'	spiral galaxy	UMa	NGC 5457, Pinwheel Galaxy
M102									There is no M102
M103	01h33.2'	+60°42'		7500	6.9	6'	open cluster	Cas	NGC 0581, C0129+604
M104	12h40.0'	-11°37'	1128	4.2E7	9.3	7' × 2'	spiral galaxy	Vir	NGC 4594, Sombrero Galaxy
M105	10h47.8'	+12°35'			9.3	2' × 2'	elliptical galaxy	Leo	NGC 3379
M106	12h19.0'	+47°18'		1.3E7	8.3	20' × 6'	spiral galaxy	CVn	NGC 4258
M107	16h32.5'	-13°03'			8.1	8'	globular cluster	Oph	NGC 6171
M108	11h11.5'	+55°40'			10.0	8' × 2'	spiral galaxy	UMa	NGC 3556
M109	11h57.6'	+53°23'			9.8	7'	spiral galaxy	UMa	NGC 3992
M110	00h40.4'	+41°41'			8.0				NGC 0205

Notes:

1. Estimates of object brightnesses and sizes often vary widely from observer to observer.
2. Sources; *The Messier Catalogue Finder card*, by Owen Gingerich, Sky Publishing Corp.; *Norton's Star Atlas*.
3. Epoch 2000 coordinates.

Index

Author's note on the index; I appreciate a good index in a book but really good indices are hard to come by. The index is basically a word association game played between the author and the reader. Fortunately, the software used to generate this book makes indexing rather painless, for I just flag every word that I think should appear here. The computer worries about page numbers and alphabetizing the list. If, however, you are looking for a reasonable entry and can't find it then write to me. I will put it in the next edition. I can be reached at: Peter L. Manly, 1533 W. 7th St, Tempe, AZ 85281-3211, USA, or via Internet at PETEMANLY@BIX.COM.